Ninety-nine Gnats, Nits, and Nibblers

Ninety-nine
Gnats, Nits, and Nibblers

May R. Berenbaum

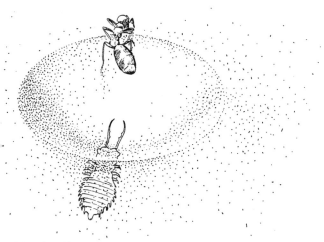

With illustrations by John Parker Sherrod

University of Illinois Press   Urbana and Chicago

© 1989 by the Board of Trustees of the University of Illinois
Manufactured in the United States of America
1 2 3 4 5 C P

This book is printed on acid-free paper.

Library of Congress Cataloging-in-Publication Data

Berenbaum, M. (May)
    Ninety-nine gnats, nits, and nibblers / May R. Berenbaum ; with
illustrations by John Sherrod.
        p.   cm.
    ISBN 0-252-01571-1 (cloth : alk. paper). ISBN 0-252-06027-X (paper : alk. paper)
    1. Insects.   I. Title.   II. Title: 99 gnats, nits, and nibblers.
QL 467.B47 1989
595.7—dc19   88-15420
                                                                        CIP

This book is dedicated to
Adrienne and Morris Berenbaum—my amazing parents.

Contents

Preface xi

Introduction xiii
A Word about Names xiii
A Word about Classification xv
The Mysteries of Metamorphosis Explained xx

1. Kitchen Companions
Common cockroaches 3
Flour beetles 6
Fruit flies 8
House flies 10
Indian meal moths 13
Mealworms 15
Moth flies 17
Pharaoh ants 19
Silverfish 21

2. House Guests
Carpenter ants 25
Carpet beetles 27
Centipedes 29
Clothes moths 31
Cluster flies 33
Cobweb spiders 35
Earwigs 37
House-dust mites 39
Termites 41

3. Garden-variety Types
Bean leaf beetles 47
Black swallowtails 49
Junebugs 51

Mexican bean beetles 53
Picnic beetles 56
Plant-feeding stink bugs 58
Spittlebugs 60
Squash bugs 62
Tobacco hornworms 64

4. Farm Friends
Alfalfa weevils 69
Blister beetles 71
Click beetles 74
Corn leaf aphids 77
European corn borers 79
Grasshoppers 81
Greenbugs 84
Green lacewings 86
Leafhoppers 88

5. Roadside Attractions
Crab spiders 93
Goldenrod ball gall flies 95
Golden tortoise beetles 97
Monarch and viceroy butterflies 99
Painted ladies 102
Parsnip webworms 104
Praying mantis 106
Sulfur butterflies 109
Woollybears 111

6. Bush Leaguers and Fruit Pickers
Apple maggot flies 115
Black vine weevils 118
Cecropia moths 120
Eastern tent caterpillars 122
Grape phylloxera 125
Gypsy moths 127
Hackberry psyllids 130
Plum curculios 132
Tiger swallowtails 135

7. Underground Gourmets
Antlions 141
Cicada killers 143

Ground beetles   145
Patent leather beetles (bess beetles)   147
Periodical cicadas   149
Rock crawlers (grylloblattids)   152
Sowbugs   154
Springtails   156
Winter stoneflies   158

8. Flexible Flyers
Carpenter and mason bees   163
Crickets   165
Dance flies   167
Fireflies   169
Hover flies   171
Katydids   173
March flies   175
Snowy tree crickets   177
Yellowjackets   179

9. The Wet and Wild Bunch
Backswimmers   183
Caddisflies   185
Diving beetles   187
Dobsonflies   190
Dragonflies and damselflies   192
Mayflies   195
Shore flies   197
Water striders   199
Whirligig beetles   201

10. Pet Peeves
Black flies   205
Bot flies   207
Cattle grubs   209
Chicken lice   211
Ear mites   213
Fleas   215
No-see-ums   218
Stable flies   220
Ticks   222

11. What's Eating You?
Bed bugs   227

Black widow spiders   230
Body lice   232
Brown recluse spiders   234
Chiggers   236
Crab lice   238
Human bot flies   240
Mosquitoes   242
Scabies   245

Select Bibliography   247
Index by Scientific Name   255

Preface

A Word of Explanation

One Saturday morning in September, 1982, Steve Sheppard, a graduate student in the Department of Entomology at the University of Illinois at Urbana-Champaign, asked me if I would like to be a guest on his radio show on WEFT-FM, a local community radio station. Steve's show, called "The Down-Home Country Show," was a mixture of old-time folk and country music, down-home recipes, fishing tips, and New Wave cracker-barrel philosophy. He thought a short program about insects would be in keeping with the theme and asked me if, as a member of the entomology faculty, I would be interested in bringing insects to the people of Champaign County (or, at least, to a few people in a portion of Champaign County, since WEFT's power allotment at the time was barely enough to reach the edge of Urbana, much less to the far reaches of the county). Ever since I became an entomologist, I've been an outspoken insect advocate, especially because, for the first decade and a half of my life, I had been terrified of anything that crawled, hopped, or flew. It wasn't until I took a college course in entomology that I realized what I was missing. In any case, I happily agreed to provide a three- or four-minute commentary every week on a different species. Steve called the feature, "Those Amazing Insects."

I confess that I was dubious at first as to the program's reception. For one thing, insects aren't among the best-loved of animals; upon hearing that I study insects for a living, most people usually say something along the lines of, "That's very nice," and then quickly change the subject. Moreover, 10:30 Saturday morning isn't exactly prime time, and even the handful of loyal listeners who never missed Steve's show generally listened while they did their vacuuming. To my surprise, however, "Those Amazing Insects" developed a following. People who knew me would stop me in the street or call me on the phone to ask about insectlike creatures they'd noticed that week. Then people who didn't know me would call or write to the station with their entomological inquiries. I began to realize that maybe people really did want to know more about insects but (as the line goes) were afraid to ask. Then, in 1987, one of the people who wrote to me about insects, Judith McCulloh, happened to work at University of Illinois Press. Not only did she listen to the show, she also thought that "Those Amazing Insects" might make an interesting book. I thought so, too, and this book is the result.

The ninety-nine species profiled here are, for the most part, subjects of radio scripts between 1982 and 1987, modified to better fit the printed page. Several species were the subjects of columns I wrote for *Horticulture* magazine, which were modified to emphasize insect biology rather than insect control. And while radio requires no illustrations, a book does, so I recruited John Sherrod at the Illinois Natural History Survey, who was not only a gifted artist but was also a fan of the program. His inimitable illustrations greatly enhance both the instructional and entertainment value of this book. Jacqueline Smith, possibly the best secretary in the Western Hemisphere, if not the world, did a phenomenal job deciphering my penmanship and producing a presentable manuscript. Finally, my friend and husband, Richard Leskosky, provided excellent (if at times painful) editorial advice and unrelenting encouragement.

One of the nicest aspects of being an entomologist is that, as a career, it offers job security unequalled in other disciplines. After all, there are almost a million species of known insects and at least twice that number as yet undiscovered. In terms of sequel potential, Rambo can't hold a candle to the class Insecta.

Introduction

A Word about Names

Every insect has at least two different names, a so-called common name and a Latin, or scientific, name. Common names are vernacular names, the words one uses to describe, for example, what just crawled by in the kitchen (e.g., "Get that filthy cockroach out of here!"). They are generally in the same language as the rest of the sentence, and they are generally familiar to the public at large. One problem with common names, however, is that, with upwards of a million different kinds of insects in the world, not all of them are common enough to have common names. Moreover, common names tend to be rather vague. The word "bug," for example, to an entomologist means a particular group of insects characterized by a distinctive set of piercing-sucking mouthparts, but to the average person a "bug" is any sort of crawling, flying, or creeping vermin with an extraordinary number of legs (or, even more broadly, anything not immediately recognizable as a bird, a fish, a snake, or a bunny).

Scientists called taxonomists are the ones who bestow upon animals their scientific names. Taxonomy is the science of identifying, naming, and classifying organisms. Scientific names are now pretty much standardized worldwide, but such was not always the case. Before the eighteenth century, scientific names often consisted of long and involved Latin descriptions of an animal (Latin being the language in which scientific research was conducted). For example, one butterfly was known as *Papilio media alis pronis praefertim interioribus maculis oblongis argenteis perbelle depictus*. The disadvantages of this system are abundantly clear; by the time somebody rushed over to tell you he'd seen one, it would be long gone before he finished his sentence.

In 1758, Carl Linné, or Carolus Linnaeus, as he was known in scientific circles (the penchant for Latinizing science even extended to scientists' names), published a book called *Systema Naturae*, in which he used a binomial, or two-name, system consistently throughout the animal kingdom. Every name had two parts; the first part, which is always capitalized, is the genus, or larger taxonomic unit, and the second name is the species, the smaller taxonomic unit. (Both names are italicized because they are Latin, and foreign languages must be italicized within English text.) This binomial system so impressed people that it was adopted throughout the world. Today, no scientific names that precede Linnaeus's book are considered valid (unless Linnaeus himself used them), and all subsequent scientific names must conform to Linnaean rules. (Those occasional three-part scientific names you may see from time to time are not real violations of the Linnaean system; rather they represent races or subspecies within a Linnaean species. A case in point is the species *Pediculus humanus*, the body louse. Body lice that live in hair on the head are known as *Pediculus humanus capitis*, and body lice that prefer to hang out in clothing or on the body are known as *Pediculus humanus corporis*.)

That the system works is evidenced by the fact that, although Linnaeus named only about two thousand insect species, there are today more than nine hundred thousand species with Linnaean names. The only people conversant with even a tiny fraction of these names, however, are entomologists. As a rule, non-entomologists have an overwhelming aversion to using, or even listening to, scientific names. However, they're not as bad as people think. For one thing, they're recognizable all over the world, so you can discuss your cockroach problems with people from all nations. For another thing, they're not as difficult as people make them out to be. Some common names are actually nothing but expropriated scientific names (or vice versa, for that matter—after all, Greek and Roman common names are the bases for many arcane-sounding scientific names). For example, the cicada, the ichneumon, and the mantis all have common names that are the same as their scientific names. Oftentimes the scientific names are identical to the common names, except they're rendered in Latin. For example, *Musca domestica* is literally Latin for "house fly," and *Pediculus humanus* translates directly from the Latin into "human louse." *Apis mellifera* is "the bee that carries honey," or (in case you haven't guessed yet) "honey bee." Moreover, scientific names aren't necessarily longer and more cumbersome than common names. The "California five-spined engraver beetle" is simply the pithy *Ips confusus*, which isn't really confusing at all. Often, when you go to the trouble to find out what the scientific names mean, they're descriptive and easy to remember. They can refer to the size, color, habits, or shape of the insect, to the locality where it can be found, or to the name of the person who first discovered it.

What this editorial is leading up to is to notify the reader that scientific names are used throughout this book. You can skip over them if you want to (I do the same with long Russian names in Tolstoy novels), but they are there for the sake of accuracy (and occasionally for the sake of amusement). If nothing else, using scientific names is a great way to impress your friends and convince them you really know what you're talking about—just ask any entomologist (they do it all the time).

A Word about Classification

Considering just how many arthropods there are on this planet, it's remarkable that the word doesn't come up in conversation very often. Far from being especially esoteric, the word "arthropod" is actually self-explanatory, at least if one is reasonably fluent in classical Greek. The word comes from the Greek words for "jointed leg," and it is the name given to the vast array of living creatures who move around by virtue of hinges in their various and sundry appendages.

There is a staggering number of arthropods in the world. According to one expert, there may be as many as 35 million species of them; according to another expert, the number of *individual* arthropods on the globe at any given time runs in the quadrillions (that's thousands of trillions, each of which is a thousand billion, for those unaccustomed to thinking big). Insect taxonomists have mercifully imposed some order on this array, using the standard hierarchical system used by all other taxonomists (table 1). Arthropods are divided into two groups based on the types of appendages surrounding their mouth. The chelicerate arthropods possess (not unexpectedly) chelicerae (pincerlike jaws) and a pair of leglike mouthparts called pedipalps. Aside from having chelicerae, the chelicerates are distinguished from other arthropods by *not* having antennae and by being divided into two main body segments: the abdomen and the cephalothorax (a head-and-chest

Table 1. The Standard Hierarchical System

Category	Example
Kingdom	Animalia (all the animals)
Phylum	Arthropoda (animals with jointed legs)
Class	Insecta (joint-legged animals with three body parts)
Order	Hymenoptera (membrane-winged insects)
Family	Apidae (bumble bees, honey bees, and a few obscure tropical bees)
Genus	*Apis* (honey bees)
Species	*mellifera* (European honey bee)
Subspecies	*ligustica* (the so-called "Italian bee," the subspecies that's used most extensively for commercial honey production in the United States)

sort of arrangement to which are attached the chelicerae and four pairs, give or take one pair, of legs). The most familiar of the chelicerates are members of the class Arachnida, a motley assortment of such unlikely animals as scorpions, whip-scorpions, wind-scorpions, and daddy-longlegs, and slightly more likely animals such as spiders, ticks, and mites. Spiders (in the order Araneida) can be distinguished from ticks and mites (in the order Acari) by size (spiders are usually larger), by shape (spiders have a narrow "waist" between the cephalothorax and abdomen, while ticks and mites have no waistline to speak of), and by feeding habits (all spiders are predaceous, while all ticks are parasitic on other animals, and mites eat anything ranging from plants to dead things to mammals to other arthropods).

The other major group of arthropods (actually, the larger of the two) consists of the mandibulate arthropods. These have mandibles—mouthparts that work together on a horizontal plane. They also have antennae (either one or two pairs) and an embarrassment of legs (any-where from three to seventy pairs). Among the lower-profile mandib-ulate arthropods are centipedes, which possess one pair of legs per body segment, and millipedes, which possess two pairs of legs per body segment. The class Crustacea, which includes aquatic forms such as crabs, lobsters, shrimps, and barnacles and terrestrial forms such as woodlice and sowbugs, have two pairs of antennae and no set number of legs (which differ from the legs of their near relatives by being divided into two parts laterally at the ends).

The group which can boast of being the dominant life form on earth, however, is Class Insecta. To recognize an insect instantly, you need only notice two things. The first is a body divided into three sections— head, thorax, and abdomen. This body division is what the class name refers to; *Insecta* is Latin for "cut into" and is in turn derived from the Greek *entomos* (look familiar?), which means the same thing. The sec-ond thing to look for is six legs, all attached to the middle, or thoracic, body segments. A few other features are often, but not always, asso-ciated with insects, wings being chief among these. Insects are the only organisms that possess wings that are not derived from other appen-dages. Birds, for example, gave up the use of two legs in order to develop appendages for flight, and bats and flying squirrels gave up the use of their arms. Insects, however, not only didn't have to give up any of their legs to fly, in some cases they managed to acquire four wings—one better than one-upmanship with respect to other flying animals.

Recognizing insects is actually easier said than done. The key char-acters (that is, characters used to define a group) may sometimes be difficult or even impossible to see. In adopting a variety of life-styles, insects have had to make a few adjustments. Many larval or immature forms (particularly larval flies, affectionately known as maggots) not only lack legs, they also lack a clearly definable head (living in manure or rotting garbage evidently doesn't require a great deal of intellectual capacity). Even professional entomologists are hard-pressed to recognize certain individuals as insects if their life cycle is not known. One example is *Microdon*, a genus of hover fly (family Syrphidae), the larvae of which hang out in ant nests in order to eat ant leavings and even an occasional ant grub. As might be expected, ants are not too thrilled with this arrangement and are prone to attack nest invaders with an arsenal of venomous stings and bites. As a result, *Microdon* larvae are tanklike, heavily armored, and almost entirely featureless, which is why for several decades they were placed in the phylum Mollusca until somebody finally figured out what the parents (who are perfectly respectable flies) looked like.

The class Insecta is further divided into groups called orders. Although entomologists, if pressed, can name over 25 different orders, most people usually encounter representatives of 15 or 16, tops. Every member of each order shares some distinguishing anatomical characteristic. In most cases, the distinctive characteristic is incorporated into the name of the order. Thus, the major orders are as follows:

Ephemeroptera (from *ephemero*, or "short-lived," and *ptera*, or "wings"). The ephemeropterans are the mayflies—soft-bodied, fragile-looking aquatic insects whose chief claim to fame, aside from two or three long tails, is the fact that adults sometimes live for only twenty-four hours.

Odonata (from the Greek for "tooth," referring to the teeth on the mandible). The odonates are the dragonflies and damselflies—large, often colorful, flying insects frequently found near water. The nymphs and adults are both predaceous. The main distinguishing feature of the odonates is the way they hold their wings—straight up over the back or out to the side rather than folded over their backs.

Orthoptera (from *ortho*, or "straight," and *ptera*, or "wings"). The orthopterans are grasshoppers, crickets, walking sticks, and katydids. Their wings fold over the body at rest. Most orthopterans are plant feeders, although some crickets are predaceous.

Dictyoptera (from *dictyo*, or "net," and *ptera*, or "wings"). Dictyopterans are the mantids and cockroaches. While some taxonomists consider these insects to be unusual orthopterans, others prefer to place them in their own order for several reasons, not the least of which include their habit of laying eggs in enclosed cases or capsules (called oothecae), their wing musculature, and the unusual three-pronged, asymmetrical genitalia of the males. While mantids are predaceous, using their enormously enlarged forelegs to snag prey, cockroaches can and will eat just about anything.

Isoptera (from *iso*, or "equal," and *ptera*, or "wings"). Isopterans are otherwise known as termites. Both the front and hind wings of termites are about the same size (hence *iso*) and are equipped with many veinlike wrinkles. Termites constitute one of the two groups of insects that live socially, with certain individuals given the responsibilities of mating and egg production and the vast majority of individuals unable to reproduce but responsible for taking care of immatures and protecting the colony from danger.

Plecoptera (from *pleco*, or "folded," and *ptera*, or "wings"). Plecopterans are commonly called stoneflies, since they are often found under or around rocks and stones in streams or on lakeshores. The scientific name refers to the way in which the hind wings fold over the back when the insect is at rest (a characteristic of interest almost exclusively to hard-core entomologists). Immature stoneflies spend their time in water, feeding on any number of different things, and, since the adults are weak fliers, they are generally found near water as well. One distinctive feature is the presence of two long jointed appendages, called cerci, on the tip of the abdomen in the adult.

Anoplura (from *anopl*, or "unarmed," and *ura*, or "tails"). Anoplurans are small bloodsucking external parasites of mammals. They're known as the sucking lice, since they draw blood with mouthparts built on the principle of a hypodermic syringe.

Mallophaga (from *mallo*, or "wool," and *phaga*, or "eat"). Mallophagans are small external parasites of both birds and mammals which chew rather than suck to obtain sustenance. Instead of feasting on blood, they more commonly feed on bits of skin, hair (like wool), or feathers (and in fact are sometimes called feather lice).

Hemiptera (from *hemi*, or "half," and *ptera*, or "wings"). Hemipterans are the true bugs. When an entomologist uses the term *bug*, he or she is referring only to insects in the order Hemiptera—that is, insects with piercing-sucking mouthparts whose wings are half-leathery (at the base) and half-membranous (at the tip). Insects that can truly be regarded as bugs include many species of aquatic bugs (including water boatmen, backswimmers, shore bugs, toad bugs, and waterscorpions), and such terrestrial forms as bed bugs, stink bugs, pirate bugs, leaf bugs, damsel bugs, assassin bugs, ambush bugs, seed bugs, lace bugs, and broadheaded bugs.

Homoptera (from *homo*, or "alike" and *ptera*, or "wings"). Homopterans are similar to the hemipterans in that they use piercing-sucking mouthparts to feed, but their wings are uniformly membranous from base to tip. Unlike the hemipterans, which use their mouthparts to suck up everything from plant juice to insect hemolymph to human blood, the homopterans are exclusively plant feeders. They include cicadas, leaf-, tree-, and planthoppers, plantlice, whiteflies, aphids, and scale insects.

Neuroptera (from *neuro*, or "nerve" and *ptera*, or "wings"). Neuropterans are delicate insects with four membranous wings crisscrossed by a nervelike network of veins. Unlike the insects in the preceding orders, neuropterans (and members of all the following orders) undergo a process called complete metamorphosis, in which immature stages (called larvae) are totally different in appearance and life-style from adults; the transition to adulthood is so extreme that it takes place during a quiescent nonfeeding stage called the pupal stage. Alderflies, dobsonflies, fishflies, lacewings, antlions, owlflies, and snakeflies are all neuropterans. Immature and adult stages are primarily predaceous.

Coleoptera (from *coleo*, or "sheath," and *ptera*, or "wings"). This order is the largest in the animal kingdom. About 40 percent of all insect species known are coleopterans, or beetles. The chief distinction of the order is the fact that the forewings, called elytra, are hardened and inflexible and usually form a protective sheath for the flightworthy hindwings. Beetles can be found doing just about anything any insect does.

Trichoptera (from *tricho*, or "hair," and *ptera*, or "wings"). Called caddisflies, trichopterans are recognized by the presence of hairs on the surface of their wings, which are held rooflike over the body at rest. Caddisfly larvae live underwater, often in self-constructed cases, and the mothlike adults tend to be found near bodies of water.

Lepidoptera (from *lepido*, or "scale," and *ptera*, or "wings"). Lepidopterans are the moths and butterflies, all of which have scales (often brilliantly colored ones) on their front and hind wings. The larvae, known to one and all as caterpillars, are almost always plant feeders, but there are exceptions to every rule, and a few rogue caterpillars actually eat other soft-bodied, plant-feeding insects.

Diptera (from *di*, meaning "two," and *ptera*, or "wings"). As their name implies, dipterans have only two functional wings. The second pair of wings is reduced to knobbed balancing organs called halteres. These are the only insects properly referred to as "flies" (as opposed to coleopterous imposters like the firefly or Spanishfly). Immature flies are often called maggots and can be found in a breathtaking variety of substrates, ranging from apples to horse intestines.

Hymenoptera (from *hymeno,* meaning "membrane," and *ptera,* meaning "wings"). Hymenopterans are the sawflies, ants, wasps, and bees. Their wings are indeed membranous, and the two pairs are joined together by means of tiny hooks called hamuli. The sawflies can be distinguished from other hymenopterans by virtue of the fact that they lack the characteristic "wasp waist," the narrowing between thorax and abdomen. The order Hymenoptera is the one other place (aside from the Isoptera) where true sociality arises in the insects, and it is thought to be widespread in the order due to peculiarities of hymenopteran genetics, whereby female insects end up more closely related to their sisters than to their own offspring and, as a result, amicably forego reproduction to raise the offspring of their royal reproductive relatives.

Although there are a few more orders that even an ardent entomophobe may someday encounter (and in fact a few are included in this book), by and large, these sixteen orders are the ones most likely to be represented in the day-to-day dealings humans have with the insect world.

The Mysteries of Metamorphosis Explained

One physical feature that distinguishes arthropods from all other organisms is their jointed external skeleton, made waterproof with a waxy outer coating and made stiff and strong by a matrix of protein and chitin (a long-chain sugar polymer). The advantages of a hard exoskeleton are numerous. For one thing, the external waterproofing prevents insects from drying out in terrestrial environments. For another thing, insect legs are basically hollow cylinders, and it's a general rule of physics that hollow tubes have greater resistance to bending and stress than solid rods made of the same amount of material. It's a smart move for insects to invest in exoskeletons because, being so small, they have only a limited amount of material to work with. From an evolutionary perspective, hardened exoskeletons also gave insects the strength and support they needed to lift their bodies up off the ground, to facilitate scurrying around as they are now wont to do. Cuticle (the external "skin") also made possible the development of wings (tough, strong yet lightweight membranous extensions of exoskeletons), and wings, of course, allowed insects to reach lofty treetops and tops of refrigerators.

Exoskeletons, however, do have certain shortcomings. Chief among these is the fact that, once you're surrounded by body armor, you can't get much bigger without bursting. As a result, all insects must molt, or periodically shed their exoskeleton, in order to grow. The process is known as ecdysis, from the Greek root meaning "getting out." For the same etymological reason, the term "ecdysiast" has come to mean "stripper." Insects approach the matter of molting basically in two ways. The simpler solution is called incomplete, or gradual, metamorphosis. In this case insects retain the same general shape and features but just get bigger with each molt; wings develop externally from wing buds and also grow progressively longer until the final molt to adulthood. One reason that molting among insects ceases upon maturity is that epidermal tissue, which synthesizes the new cuticle to replace the exoskeleton that splits and is shed, degenerates in wings and can no longer produce new cuticle. Collembolans (or springtails) and thysanurans (or silverfish) lack wings and continue to molt throughout adulthood. Orders with gradual metamorphosis include the Orthoptera, Hemiptera, Homoptera, Dictyoptera, Mallophaga, and Anoplura. Each developmental stage is called an instar, and the immature forms of insects with gradual metamorphosis are known as nymphs.

A somewhat more complex solution to metamorphosis was adopted by the so-called higher orders, including the Hymenoptera, Lepidoptera, Diptera, and Coleoptera. Development in these orders is characterized by a resting (or pupal) stage, during which virtually all body tissues break down and are reorganized. As a result, immature insects with complete development, called larvae, tend to look radically different from their parents. Maggots don't resemble flies, caterpillars don't resemble moths, and grubs don't resemble beetles. Complete metamorphosis, although thought by some to be more recently evolved than gradual metamorphosis, has been a big hit; the vast majority of insects on earth use this form of development. One possible reason for its great success is that insects with complete development can exploit very different habitats throughout their lives—larvae can live in places adults wouldn't be caught dead in and vice versa.

Chapter 1

Kitchen Companions

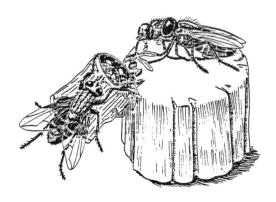

Common cockroaches

Flour beetles

Fruit flies (*Drosophila*)

House flies

Indian meal moths

Mealworms

Moth flies

Pharaoh ants

Silverfish

COMMON COCKROACHES

Fossil remains dating back to the Carboniferous Era, about 300 million years ago, contain in some places, to the exclusion of just about everything else, the remains of insects—cockroaches, to be specific. Those ancient roaches are virtually identical to their contemporary counterparts. The Carboniferous Era, in fact, is often referred to by the entomologically inclined as the Age of Cockroaches, however difficult it may be for some people to believe that there was ever an era with more cockroaches than nowadays.

It's likely that everyone has, at one time or another, cohabited with cockroaches. Although there are over fifty species of cockroaches in North America, the vast majority are inoffensive denizens of woods and forests. There are four, however, that live almost exclusively in the company of humans. About the most widespread among these uninvited household guests is the German cockroach, *Blatella germanica*. Despite its name, the German cockroach is cosmopolitan in distribution and is in fact called the Russian roach by the Germans. The German roach is no doubt the most ubiquitous household pest and is a familiar if unwelcome sight, averaging a half-inch in length and pale brown in color, with two dark parallel stripes on a shield behind its head.

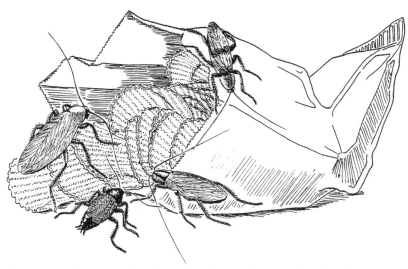

Common cockroaches. The diners from left to right are: the American roach, the Oriental roach, the German roach, and the brown-banded roach.

3

Say what you will about it, one must admire its powers of reproduction: the German roach breeds year round, irrespective of season. And unlike most insects, the female doesn't merely scatter her eggs at random and simply forget them; she cements about thirty to forty-eight eggs together into a hard-shelled capsule which she carries around with her until just before the eggs hatch (in about three weeks). This sort of parental care, even if minimal, guarantees that virtually all the eggs will hatch, provided that Mother doesn't inadvertently stray out onto a well-illuminated wall and get squashed unceremoniously along with her progeny. One female can produce about five capsules (called oothecae by those in the know) in a lifetime, and, about three months after hatching, her female offspring are ready, willing, and able to do the same. Left alone, it's theoretically possible for one female German roach to produce over thirty thousand offspring in a single year. In fact, there may be places in your town where this is going on right now.

The Oriental cockroach, *Blatta orientalis*, is far more conservative than its German relation. Among other things, it mostly frequents basements and other dark protected areas. Oriental roaches are black or dark brown, 1 to 1¼ inches in length, and are sluggish compared to most other roaches. The females are virtually wingless. (While most roaches are no great aerialists, they can at least manage to flutter off a table should the occasion arise.) The Oriental roach is perhaps best noted for its peregrinations. It freely walks from building to building, particularly in the fall, to infest new basements through drains, cracks in walls, and unopened doors.

Not all cockroaches have garnered exotic place names; there is an American cockroach, too. *Periplaneta americana*, the American cockroach, outdoes the Oriental in size, and it can reach lengths of two inches or more. American cockroaches are reddish brown to dark brown and are partial to basements and sewers. Because of this predilection they (and other cockroaches) are often called waterbugs. They're also more conservative reproductively than most cockroaches, producing only twelve eggs in a capsule, which they glue to objects in a protected place rather than carry about. The nymphs (or immatures) take up to a year to mature. Name notwithstanding, American cockroaches are thought by some to have originated in Africa and to have arrived in the U.S. on slave ships several hundred years ago.

The relative newcomer to the U.S. indoor cockroach community is the brown-banded or TV roach, *Supella longipalpa*. This roach made inroads during World War II when GIs returned from action in the Pacific theater and brought them home inadvertently as souvenirs in household goods. They're smaller than Germans, topping off at half

4

an inch, and have longer antennae and a characteristic light tan band across the base of their wings. They prefer warmth and dryness to cold and damp and are gaining prominence in houses with central heating. They actively seek out warm places, and they're the roaches one commonly finds inside TV cabinets (hence the name), in alarm clocks, and in among the coffee dregs in percolators. They reach maturity in less than two months and rival Germans when it comes to reproductive effort. As for cockroach control, sanitation is still probably the best method. Fix dripping sinks and sweep up crumbs (no sense, after all, feeding and watering them). Insecticide resistance is a serious problem, particularly in the German cockroach, which has not only developed biochemical resistance to most formulations but has learned to avoid places where insecticides and dusts have been used. Boric acid and sticky traps are probably the least nasty and most effective insect killers on the market today for home use, pesticide applicators notwithstanding. It's highly improbable that any pesticide will remain effective for long or that eradication will be complete, however. The fact that cockroaches have done just fine for the last 300 million years does not bode well for us for the next few centuries.

FLOUR BEETLES

Flour beetles can be found year round as permanent residents of any place where grain or food products are stored. In one sense they're ideal house guests, since there are very few foods not to their liking. They're equally at home feasting on any grains, starchy materials such as dried crackers or pasta, peas, beans, dried plant roots, chocolate, dried fruits, nuts, snuff, cayenne pepper, drugs, and, of course, flour. Flour beetles belong to the genus *Tribolium*, in the family Tenebrionidae, and they're among the tiniest members of the family. The largest *Tribolium*, *T. giganteum*, is something of a misnomer at only 10 mm in length. The smallest, *T. castaneum*, bottoms out at only 2.5 mm (about 1/10 inch) in length.

The most common species of *Tribolium* worldwide are *T. castaneum*, the red flour beetle, and *T. confusum*, the confused flour beetle. *T. confusum* is not so much confused as it is confusing, at least to entomologists called upon to distinguish it from the very similar *T. castaneum*. The differences are, to say the least, on the arcane side. The red flour beetle has a distinct club at the end of its antennae (as opposed to a gradually enlarged tip), its eyes are a bit closer together, and its head lacks the characteristic notch displayed by the confused flour beetle.

Needless to say, none of these entomological niceties are of much interest to the homemaker who discovers things crawling in his or her flour. The two species share pretty much the same life-style. Neither species requires any drinking water at all to survive; they're perfectly

Flour beetles

capable of surviving at zero percent relative humidity, extracting all the water they need from carbohydrates in their food. Females lay eggs in whatever is on the menu at the moment, and, after three to twelve days under normal household conditions, the eggs hatch into six-legged, brownish-white grubs (sometimes called bran bugs). The length of the larval period depends largely on the availability and quality of food; grubs may molt anywhere from five to eleven times and take from three weeks to three months to pupate. The larvae tunnel to the surface of whatever they're feeding on to form naked white pupae and emerge as adults one to two weeks later. As insects go, they're fairly long-lived as adults, and they continue to reproduce for most of their adult life. If she lives a clean life, a female living a year or so can lay up to a thousand eggs.

Which is not to say *Tribolium* is known for clean living. They have a number of habits that in other organisms would be considered reprehensible, to say the least. When conditions get crowded, for example, and food is in short supply, adult beetles don't hesitate to eat their own offspring, concentrating on the immobile and defenseless pupae and eggs. Male confused flour beetles routinely mate with anything crossing their path that remotely resembles a flour beetle, including red flour beetles of either sex, often causing fatal injury to the reluctant partner before discovering the mistake. And both sexes excrete vile-smelling chemicals called quinones as they feed, which discolor and contaminate even the flour they don't eat.

Vile though they may be, flour beetles have been constant companions of humans for millenia—one species was found in a pharaoh's tomb of the sixth dynasty dating back 4,500 years. The fact that flour beetles are almost never found except in the company of humans has given rise to much speculation as to what they did for a living before humans were around. Most *Tribolium* species survive better on crushed or particulate grain than on whole grain. A great moment for *Tribolium*, then, was when the process of milling was invented—not a "run-of-the-mill" discovery from the flour beetle point of view.

FRUIT FLIES (*DROSOPHILA*)

New Year's Eve is traditionally ushered in with convivial toasts and free-flowing spirits—and New Year's Day is traditionally spent suffering the consequences. *Drosophila melanogaster*, however, ushers in each new day with an excess of alcohol, and no tippler can hold his liquor any better. *D. melanogaster*, a tiny yellowish stout-bodied fly about one-tenth of an inch long, is better known as the common garden-variety fruit fly, a familiar if unwelcome sight anywhere decaying fruit is found. As a member of the family Drosophilidae, it is more properly known to discriminating entomologists as a vinegar or pomace fly ("fruit fly" having been preempted by the apple maggot fly and friends in the family Tephritidae). *D. melanogaster* is found wherever fruit ferments, which pretty much covers the entire globe. One renowned biologist refers to them as "animal weeds." Occasionally *D. melanogaster*

Fruit flies

oversteps the fine line between a mere nuisance and an economically important pest; it causes problems in tomato processing plants and is particularly pestiferous in wineries worldwide. Not only can *D. melanogaster* hold its liquor, it can actually digest it as well, at concentrations in excess of 9 percent. One study documented that fruit flies can even flourish on just fumes of alcohol. Normally, ethanol (the alcohol that puts the kick in liquor) is a highly toxic substance, acting as a systemic poison in most living organisms. However, *D. melanogaster* produces an enzyme called alcohol dehydrogenase (ADH). This enzyme facilitates the conversion of toxic alcohols to more manageable aldehydes and ketones. While most organisms are capable of producing the enzyme, *D. melanogaster* is particularly well endowed. ADH activity can be found throughout its entire body, even (for reasons unknown to science) in the male genital apparatus. *Drosophila* larvae, the cream-colored maggots found in infested fruit, and even its eggs, produce ADH in abundance, a happy situation for female flies who deposit their eggs in grape must or wine cellars where ethanol is present at concentrations of 19 percent or greater.

D. melanogaster takes no fermenting fruit before its time—that is, while it can live on alcohol, it prefers to feed on the yeasts that produce the alcohol, and growth and development are maximal under conditions in which yeasts thrive. But preferences notwithstanding, the secret to *Drosophila's* success is its flexibility. If yeasts aren't available, then fungi, slime fluxes, and bacteria suffice. Of course, one of the readiest sources of rotting things is human beings, and as a result *D. melanogaster* is a constant associate of humans.

Conspicuousness has its price, however—*Drosophila's* reproductive habits (a generation time just under two weeks and a reproductive rate of about two thousand eggs per female), its small size, and its general willingness to feed on rotting garbage have led to its widespread domestication and use in genetics laboratories. Literally millions of fruit flies are etherized each year in the name of freshman biology. Such, however, is the price of an easy meal. *Drosophila melanogaster* will no doubt continue to be a constant companion to humans for this New Year and for New Years to come, and in the words of one forgotten sage, "Though time flies like an arrow, fruit flies like an apple"—a rotting apple, that is.

HOUSE FLIES

On the all-time hit-parade list of vermin, flies are definitely in the top ten. Flies have literally plagued humanity for centuries; at least two, and as many as four, of the biblical plagues are directly or indirectly attributable to flies of one sort or another. Chief among the two-winged offenders is the house fly, *Musca domestica*. It's one of the most widely distributed insects and one of those most closely associated with humans. The house fly is instantly recognizable to most, with its gray thorax striped longitudinally with four black lines. Everyone has at one point or another shared company with a house fly.

It's not so much for companionship that the house fly seeks us humans out, but more for our garbage. Flies are amazingly unfussy; sewage, food waste, excrement, dead animals, and other organic offal are all called home with aplomb. They are partial, however, to horse manure. Once, a half-ton pile of manure, left exposed for only four days, was found to contain four hundred thousand fly larvae (averaging four hundred flies/pound).

Female flies lay their eggs on any warm rotting material in batches of twenty-five to one hundred at a time. Each female can lay up to a half-dozen batches. The eggs develop amazingly rapidly, even for insects, and hatch within twelve to twenty-four hours. The maggots are fairly featureless white worms, legless and eyeless, with two pointed hooks at what would be the head end if they had a recognizable head. Burrowing through garbage, maggots can complete development in less than two weeks if conditions are right.

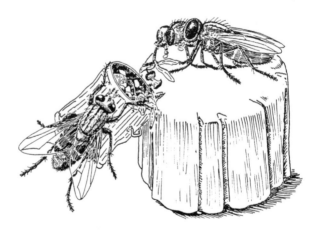

House flies: "And now for the pièce de resistánce . . . Voilà! Garbage!"

This sort of lightning-fast development has given rise to the kind of calculations entomologists are invariably enamored of. In 1906, L. O. Howard began to calculate the maximum number of offspring resulting from a single female in one season in Washington, D.C., given the unlikely probability that all survived, and reached 5 trillion 598 billion 720 million by September 15. Another individual estimated that this number could cover the earth to a depth of forty-seven feet. These figures were disputed in 1964 by noted fly expert Harold Oldroyd, who remarked that "a layer of such thickness would cover only an area the size of Germany, but that is still a lot of flies."

If flies would stick to garbage, they wouldn't be nearly as much trouble as they are, but it's their habit of frequenting both sides of the tracks that causes problems. The mouthparts of the adult house fly work on the principle of a sponge, so flies are restricted to liquid food. If something is particularly delectable yet not liquid, flies regurgitate saliva to liquify it and sop it up. So flies go directly from feeding on such delights as human excrement or garbage to open sugar bowls on tables, bringing with them, and spewing out, an indescribable variety of bacteria. By one count, flies collected in a slum district contained an average of 3,683,000 bacteria (with one particularly industrious individual transporting over 6 million). Not just bacteria are delivered from human waste to the dinner table; flies have been implicated in the spread of hookworm, tapeworms, and pinworms, over and above typhoid, dysentery, and diarrhea. They're aesthetic annoyances as well, generating fly specks on doors, walls, and windows. The light-colored specks are regurgitated food, and the dark ones, fly feces (in the event that anyone is curious).

That the earth is not forty-seven feet deep in flies suggests that flies have their problems surviving, too. The replacement of the horse by automobile made great inroads in reducing fly problems by reducing the availability of one of their favorite foods, horse manure. (In retrospect, it's ironic that the automobile was heralded as an environmentally safe, nonpolluting form of transportation in its early years for this reason.) Flies are also subject to an epidemic fungal disease that is responsible for the occasional sight of dead flies awkwardly projecting perpendicularly from ceilings or windows. Fungal spores projecting from their mouths cause the fly carcasses to stick to surfaces. There are also parasitic wasps that search diligently through manure piles for fly puparia (the resting stage between maggothood and adulthood consisting of a regular pupa encased in a hardened larval skin) and lay eggs that develop inside the fly.

But humans, of course, are the fly's greatest enemy as well as provider. Properly disposing of waste deprives them of food, and swatting and spraying take their toll. Never hesitate to kill a fly; it's a public service. Above all, don't get sucked in by its small size and innocuous appearance, as did William Oldys, an eighteenth-century poet who wrote "On a fly drinking from his cup":

> Busy, curious, thirsty fly,
> Drink with me, and drink as I.
> Freely welcome to my cup,
> Couldst thou sip and sip it up;
> Make the most of life you may,
> Life is short and wears away.

No doubt Oldys didn't realize that sharing a beer with a fly is one great way to risk shortening your life via typhoid, dysentery, or other insect social diseases.

INDIAN MEAL MOTHS

Budding entomologists and other admirers of insects need not despair if the weather is damp, cold, and dismal—just because insects aren't flying outdoors at the moment doesn't mean they're not on the wing indoors. One conspicuous element of the indoor fauna is the Indian meal moth, *Plodia interpunctella*. The Indian meal moth is a tiny little moth about three-fourths of an inch long that appears without invitation or explanation mostly on walls and ceilings of kitchens intermittently thoughout the year. It is distinctively marked, as small brown moths go. The base of its wings is silvery white with a coppery sheen, so the moths are even color-coordinated with many kitchen appliances.

The Indian meal moth doesn't materialize out of thin air, however. It actually spends its formative weeks in any kind of dried food—literally from soup to nuts (as long as the soup is dehydrated). Larval

Indian meal moths: "Joe . . . Joe, you in here?"

development takes about eight weeks, and the mature larvae are generally whitish with a green or pinkish hue. In flour they spin silken threads as they feed, and in commercial flour operations they can cause tremendous problems by clogging machinery with their webbing. At home, they mostly gross people out. They call attention to themselves when they spin a cocoon to pupate, at which time they tunnel through whatever they're feeding in to spin cocoons on the surface. In the course of their perambulations they cake the flour, leaving mold and excrement in their wake. Once out of their medium in search of a nice corner to pupate, they often engage in the unsettling practice of walking along kitchen cabinets and on walls and ceilings in full view of the horrified homemaker.

The Indian meal moth is one of many stored-product pests which humans actively recruited into existence by obligingly piling up large amounts of grain and other food for storage in Neolithic days. Stored grain provides all the essential nutrients for a developing insect in an exceedingly amenable environment boasting of constant temperatures, no major predators, and an unlimited food supply. About the only problem faced by stored-grain pests is low water availability, but many stored-product insects have even managed to get around this problem. For example, the Mediterranean flour moth, *Anagasta kuhniella*, another moth that frequents flours and cereals, can grow on foods with less than 1 percent moisture; the moisture content of most fruits and vegetables is on the order of 80 to 90 percent.

In granaries and flour mills, the Indian meal moth and Mediterranean flour moth cause considerable damage by eating the seed germ. Moreover, they foul the meal, encouraging the growth of mold and fungi. Spoilage, however, puts a limit on the growth of these guys. When grain begins to rot, it heats up due to the heat of fermentation, up to 106°F even in winter. This accelerates the growth of mold to such an extent that the flour is rendered unsuitable for moths and humans alike.

About the only practical control measure at home for stored-grain pests is simply to toss the contaminated product. If you choose not to waste what you've got, freezing the whole works for several days will kill all developing eggs, larvae, and pupae. You run the risk, then, of occasional insect parts in cookies, cakes, and other baked goods, but that's the norm in processed food anyway. The U.S. government sets standards for the number of insect parts in various foodstuffs—a quick calculation puts the estimated number of insect parts in a peanut butter and jelly sandwich at about fifty-six. If that thought's not enough to put butterflies in your stomach, just go eat a peanut butter and jelly sandwich.

MEALWORMS

When Harry S. Truman uttered the immortal words, "If you can't stand the heat, get out of the kitchen," he was obviously not addressing his remarks to the Tenebrionidae of the world. Not only are there tenebrionids that relish the heat, there are those that make themselves at home in kitchens on all continents.

Tenebrionids are variously known as mealworm, flour, darkling, or false wireworm beetles. Species in the Middle East have some remarkable adaptations for keeping cool in the hot dry deserts of the region. Those that are active during the day are generally long-legged; by standing on tiptoe (or tip-tarsi) they can raise themselves a half-inch or more above the ground surface, where the temperatures are at least twenty degrees cooler. Others have hairy bristles on their hind legs to form "sand shoes" which they use to "swim" through the sand. Many diurnal species wear white, after the fashion of the desert bed-

Mealworms

15

ouins; the white color arises from the presence of fine airspaces in the cuticle. Many do not require drinking water, although there is one that does like to obtain water by climbing up sand dunes early in the morning and standing on its head, thus allowing fog to condense on the tip of its abdomen and run down toward its mouth.

One of the most familiar tenebrionids, *Tenebrio molitor*, is a denizen of dining rooms rather than deserts. Called the yellow mealworm beetle, *T. molitor* lives and breeds in grain and coarse cereal, particularly in sacks, bins, or bags that are slightly moist. Despite its proclivity for moist meals, the mealworm is remarkably tolerant of dry conditions and can even absorb water directly from the air at humidities lower than 50 percent. Adult mealworm beetles are shiny black or brown, from a half-inch to an inch in length.

Females lay up to five hundred to one thousand bean-shaped white eggs in flour or meal. The eggs are sticky and are quickly coated with flour. After two weeks, slender cylindrical white larvae hatch and gradually turn darker. Early on, they're yellow, and when they reach full size they turn yellowish brown, particularly at the ends. (A related species, *T. obscurus*, is known as the dark mealworm, since its larvae are—guess what—much darker). Larvae reach full size in about three months but remain in the larval stage through the winter. In spring, pupation lasts, on average, about two weeks. Depending on conditions, then, a generation can take anywhere from four months to two years. The pupae are curiously vulnerable; they are not covered by a hard shell or cocoon but are instead soft, white, and exposed to the elements. About their only protection lies in the gin-traps, toothlike extensions of the mealworm abdominal wall that can snap together like a steel trap (or, for those familiar with the cotton business, like a cotton gin trap, although parasitic wasps or mites are the only things likely to get trapped in them).

Mealworms are no doubt most familiar to people as fodder for various sorts of captive animals. Pet stores raise them and sell them as food for everything from turtles to toads, fish, and birds. But it's not surprising that *Tenebrio molitor* is so widely used to feed pets and laboratory animals; after all, everybody knows that at the heart of good nutrition is three square "mealworms" a day.

MOTH FLIES

When faced with the alternative "sink or swim," moth flies (flies in the family Psychodidae) do both. Actually, to be perfectly accurate, they swim in sinks—*and* bathtubs, basements, drains, toilets, and anywhere else raw sewage might be available. Moth flies, also known as drain flies, filter flies, or sewage flies, can reach epidemic proportions in homes, emerging from sinks, washbasins, and drains. They are quite distinctive; they're tiny, less than one-tenth of an inch long, and hold their wings rooflike over their backs when they're resting. They owe their euphemistic moniker "moth fly" to the fact that, unlike most flies, their wings are covered with a dense coat of scalelike hairs.

Female moth flies lay their eggs in clusters of thirty to one hundred on the surface film of sewage, whether in filter beds in lagoons or in drains or grease traps in basements. The tiny maggots live in moist organic material and feed on algae, bacteria, fungi, or just plain filth. They develop quite rapidly and emerge as adults only three to four weeks after the eggs hatch.

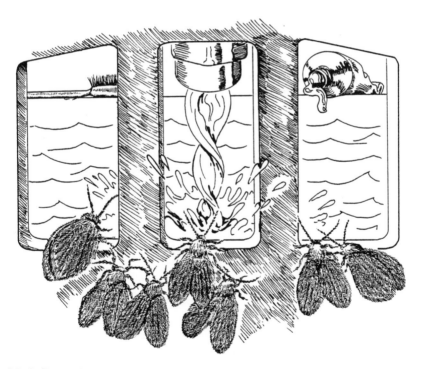

Moth flies: "I hope he turns the water off before it reaches the overflow hole."

Among the more common species of moth fly is *Psychoda alternata*, sometimes called the trickling filter fly. This fly is not dangerous but can make a total nuisance of itself by virtue of sheer numbers. Hundreds can enter a home, either through open windows overlooking a sewage lagoon or through open drains. They are weak fliers but nimble on their feet, and they have the annoying habit of crawling conspicuously all over walls, countertops, and ceilings.

While moth flies are simply annoying, their closest relatives, the sand flies (in the subfamily Phlebotominae), are downright dangerous. The sand flies begin life in the same sort of humble surroundings as the moth flies, although they needn't live in liquid waste—semisolid dung and decaying debris suffice nicely. After four to six weeks, the adults emerge. They are far less hairy than the moth flies and their wings are held up and out at about a sixty-degree angle from the body.

The main difference, however, is that adult female sand flies suck blood. While many species content themselves with the blood of lizards, frogs, and snakes, a few specialize on species of warm-blooded mammals, humans included. While the males lack the piercing mouthparts of the females, they nonetheless visit humans to dine on sweat, tears, and other noninvasive secretions. The blood-feeding habits of sand flies, combined with their love of filth, bestow upon them a talent otherwise lacking in the family—they are remarkably efficient at transmitting disease. The list of serious diseases vectored by phlebotomine flies includes oroya fever, verruga peruana, sand fly fever, kala-azar (or dumdum fever), uta, oriental sore (or Baghdad boil), chiclero's ulcer, and espundia. The reason that these diseases (caused primarily by parasitic protozoans) are not exactly household words in the U.S. is because the medically important species of sand fly are confined, at least in the Western Hemisphere, to the tropics and subtropics. So next time you complain about winter weather, think of it as a preventive against dumdum fever and Baghdad boil.

PHARAOH ANTS

If you've been feeling inexplicably antisocial lately, look around—you may be sharing your domicile with several thousand pharaoh ants. The pharaoh ant, *Monomorium pharaonis*, is one of the most persistent and pestiferous ants to invade the sanctity of hearth and home. Although they've been known to nest in soil, they much prefer to invade houses, where they can set up residence any place that an insect only one-twelfth of an inch long can gain access—in other words, in furniture, between floors, behind baseboards, in garbage, in boxes, inside foundation spaces, in stacks of papers, and even in linens and clothing. Ensconced in such comfortable quarters, pharaoh ants can live and breed year round, even in cold climates.

The tiny reddish yellow workers of the colony are not conspicuous when they leave their place of residence and forage for food. There is essentially nothing that humans eat that pharaoh ants won't deign to dine on; they dote on sweets, bread, greasy food, and also meats and other high-protein food—effectively, everything that finds its way to the table. Moreover, they eat things humans wouldn't rightly consider edible, including dead and living insects (including bed bugs) and even bathroom sponges. To top it all off, the foraging workers have an extremely efficient communications system. Scouts leave the colony at rates of about five each minute, and when one finds a promising mother lode, two hundred workers will be on site within only ten minutes.

Expediency is the slogan by which the pharaoh ant lives and is one reason that the species is so alarmingly successful. They live just about anywhere and eat just about anything. Even their social life has the

Pharaoh ants

air of expediency. For most ant species, when a colony becomes unmanageably large, new queens (fully reproductive winged females) are produced who then fly off to mate and found a new colony. Pharaoh ants, however, don't take the time to raise up new queens. A group of about fifty workers sets off on foot carrying about as many immature grubs with them, until they hit upon a likely spot to make camp. They then raise up the brood they've brought with them. Even though winged queens may be among the offspring, these queens don't leave the colony for the typical ant nuptial or mating flight. They stay in the nest and lay unfertilized eggs (which become male offspring) and eventually mate with one of their own sons.

One other aspect of pharaoh ant social life that contributes to their stunning success is the structure of the colony. Most ant colonies have only a handful of functionally reproductive individuals or queens; the vast majority of individuals are sterile female workers who take care of the daily chores. In a pharaoh ant colony, however, there may be as many as one hundred queens per two thousand workers. With 5 percent of the population busy producing eggs, it's not altogether surprising to find yourself suddenly sharing your kitchen, bedroom, bathroom, and closets with pharaoh ants.

The pharaoh ant hails originally from Africa and has managed to establish itself as a world-class pest due to the inadvertent assistance of humans traveling all over the globe. Such small insects with such unfussy habits make for excellent stowaways, and in fact pharaoh ants are often referred to as "tramp ants." What with their social structure (where one out of every twenty is royalty) and their predilection for garbage of all descriptions, it's unlikely that King Tut would have approved of them, name notwithstanding.

SILVERFISH

We've all had to eat our words from time to time, but with silverfish it's a regular occurrence. Silverfish are silvery cigar-shaped insect inhabitants of bookcases, shelves, walls, basements, porches, and other nooks, crannies, and corners of human habitations. They are easily recognized by the two long antennae on their heads and the three equally long tails on their opposite end. The tails, in fact, give rise to the name of the order Thysanura—Greek for "tassel tail." The Thysanura have a consuming fondness for carbohydrates and devour them in any way, shape, or form. These range from such traditional fare as starch-rich flour and oatmeal to more unorthodox comestibles such as wallpaper paste, bookbindings, card files, rayon, lace curtains, and paper with glue or paste as binding. They'll even eat the starch out of the starched collars and cuffs of formal dressware.

As insects go, thysanurans are about as primitive as an arthropod can get and still be considered an insect. They have no wings, for example—one feature that sets their kin apart from the rest of the insects. In addition, while most insects undergo development via a series of molts that cease at adulthood, thysanurans continue to molt after sexual maturity. They do so with great fervor—one individual was recorded to have undergone over forty molts in its 2½-year life span. Even their sexual practices are primitive by insect standards. The thy-

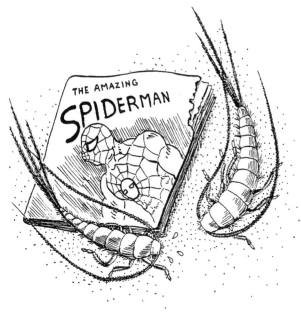

Silverfish

21

sanurans never developed internal fertilization, so the process of pairing up is less than precise at best. When a male is in the presence of females, he drops his sperm, packaged in a structure called a spermatophore, on the ground. He then grabs the nearest accommodating female and, using thin strands of silk, straps her down in the vicinity of the spermatophore. Left with little recourse, the female then picks up the spermatophore and inserts it into her genital opening.

Crude as it appears, the process must be effective enough, since thysanurans are prominent pests in many homes. Two species are the most frequent uninvited guests. The silverfish, or slicker, *Lepisma saccharina*, is about a half-inch long and silvery gray in color. The silver color is due to the presence of a dense covering of scales. The scales account for the "fish" part of "silverfish" and the occasional appellation of "fishmoth." Silverfish are nocturnal and secretive and can move very quickly out of sight when disturbed. Egglaying's about as casual an affair as mating. Females simply drop their eggs as they crawl in among cracks, crevices, and crannies. When the eggs hatch, the immature silverfish look like miniature adults and undergo no major physical changes during the course of development, which can take two years and a lot of back issues of *National Geographic* to complete.

Silverfish like places that are warm and damp and thus frequent bathrooms, baseboards, closets, and cupboards. Their relative, the firebrat, *Thermobia domestica*, prefers things hot and dark, so it's more commonly found around furnaces, in heating pipes, and in bakeries and factories. Though more mottled in appearance than the silverfish, the firebrat is basically similar in its biology.

Lest humans feel unduly persecuted, it should be mentioned in fairness that silverfish are uninvited guests in the homes of other species, too. Some species of silverfish live inside the nests of ants and termites where they eat not the books, clothes, and personal belongings of their reluctant hosts, but their *children*. So next time you find a silverfish making short work of a paperback novel on your bookshelf, be grateful for small favors.

Chapter **2**

House Guests

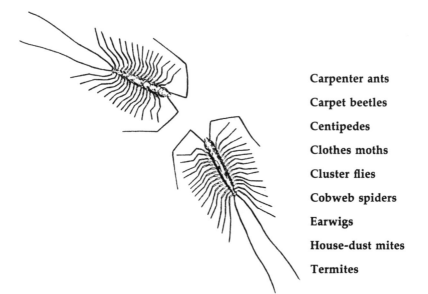

Carpenter ants

Carpet beetles

Centipedes

Clothes moths

Cluster flies

Cobweb spiders

Earwigs

House-dust mites

Termites

CARPENTER ANTS

Especially in spring, the term "antiestablishment" takes on other than political connotations; otherwise respectable houses soon become host to all manner of ants. Of these, *Camponotus pennsylvanicus*, the carpenter ant, is perhaps the most conspicuous, if by no other virtue than sheer size. These chunky black ants top off at around one-half inch in length (they were once known as *Camponotus herculeanus*). Like all species of ants, carpenter ants have a constricted hourglass waist, elbowed antennae, and front wings substantially larger than their hind wings—characteristics that distinguish them from the thick-waisted termites with their straight antennae and even-sized wings. Although termites are often called white ants, they're not even distantly related to the true ants, who, like *Camponotus*, belong to the family Formicidae in the order Hymenoptera (along with bees, wasps, sawflies, and other family members).

Carpenter ants do, however, share at least one annoying habit with termites. They construct extensive nests in wood, including logs, stumps, tree trunks, telephone poles, and, unfortunately, the timbers, joists, and windowsills of buildings. While termites actually eat and digest wood, carpenter ants simply chew and tunnel through it to build their

Carpenter ants

25

homes (hence the name "carpenter ants," although few if any human carpenters regularly masticate their handiwork). The end result, though, is much the same. Damage by carpenter ants can leave household structural timber open to fungus, rots, and other forms of decay.

Although carpenter ants have little respect for human structures, their own social life has a structure that's quite complex and includes such admirable traits as collective care of the young and division of labor. The vast majority of ants in any colony (the ones most likely to be crawling though your kitchen cabinets) are workers, whose responsibilities include gathering food, raising the young, and protecting the colony. Generally every colony has a queen, a greatly enlarged female that is capable of laying thousands of eggs over the course of a lifetime. The good news is that most colonies have only one, or a very few, queens; the bad news is that carpenter ant queens can live for ten years or more.

When a colony gets too large, it splits, and winged males and females develop and take off to mate and find new colonies. Strictly speaking, only the female founds a new colony—the male dies shortly after mating, while the new queen loses her wings and heads for a likely-looking hunk of wood to begin the business of egglaying in earnest.

This mass-mating or swarming behavior of ants is seasonal, and, for carpenter ants, spring is swarming time. Swarming is when carpenter ants first become conspicuous in houses. Ants also invade houses in spring because they're hungry. Carpenter ants are among those species of ants with an incurable sweet tooth (or more appropriately, sweet mandible). They raid houses in early spring looking for sweets, because one of their normal sources of sugar—the sweet honeydew secretion of sap-feeding aphids—isn't available until the weather warms up.

Their love of sweets has on numerous occasions proven their downfall. One of the most effective ways to control carpenter ants is to set out poison baits. Attracted to the sweet taste, the worker ants collect the bait and bring it back to the colony, where they altruistically share it with the developing larvae and the queen. Which just goes to show that there's something after all to that old saw "Too many sweets aren't good for you."

CARPET BEETLES

In the cold winter months there's nothing better than being snug as a bug in a rug—that is, unless it's being snug as a *beetle* in a rug. There are at least four species of beetles, all in the family Dermestidae, that take this sentiment to heart. Adults of *Attagenus megatoma*, the black carpet beetle, spend the warm summer months outdoors visiting flowers and eating pollen. When winter arrives they move indoors for cozier accommodations. Females lay up to one hundred eggs in various places, particularly where lint accumulates (as in air ducts, underneath baseboards, or under furniture). The eggs hatch in about a week; after egg hatch, things get less precise. The larval stage of the black carpet beetle can last anywhere from nine months to three years, depending on the availability of food. Food to these guys is a matter of opinion. They willingly feed on cereals, dead insects, dust, or animal droppings, but they're particularly fond of fur or feathers and have earned themselves the enmity of humanity by eating holes in clothing, upholstery covers, curtains, furniture stuffing, and (especially) rugs or carpets (hence the name, carpet beetle).

Carpet beetles are exceedingly well equipped for their dietary habits. Most insects (and other animals with pretenses about discriminating taste) avoid animal hair or feathers as food since they contain keratin, a relatively indigestible form of protein characterized by refractory disulfide bonds. The carpet beetles are relatively unique in that they are

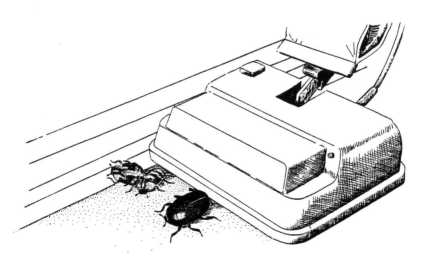

Carpet beetles: "Run! It's the vacuum cleaner!"

among only a handful of animals (including the clothes moth) that can produce enzymes that break disulfide bonds and metabolize keratin. In other words, they can literally split hairs.

Aside from the black carpet beetle, other dermestids that can be found in rugs or furniture are the common carpet beetle, *Anthrenus scrophulariae*, and the varied (or furniture) carpet beetle, *Anthrenus verbasci*. None of the carpet beetles ever exceeds one quarter-inch in length at maturity. The common carpet beetle is about one-eighth of an inch long and is mottled black-and-white with red markings along the underside. The furniture carpet beetle is black with yellowish-whitish zigzags, and the black carpet beetle is black or dark reddish-brown. All are scaly or hairy, and even the larvae are hairy (so "wearing a rug" seems only an affectation). The larvae of *Anthrenus* are oval with brownish or black hairs and six tufts of bristles on their tail; the larvae of *Attagenus* are more carrotlike in shape.

Larvae molt from six to ten times or more and leave their shed skins lying around to dismay houseowners. The skins are easier to find than the larvae, which like to hang out in cracks and crevices and avoid the light. Pupation generally occurs from April through June and can be a month-long process. The relatively unprotected pupa rests inside the last larval skin (as does the newly molted adult) for up to several weeks. Females generally mate and lay eggs in the house before heading outdoors for a few weeks of pollen feeding and flower visiting.

Carpet beetles can be serious household pests and are difficult to control once an infestation sets in. Prevention, as good an approach as any, consists of removing lint and dust and eliminating cracks and crevices in floors and baseboards. A full-scale infestation, however, is up to a professional to control. After all, carpet beetles are quite accomplished insects—especially in view of the fact that they can cut a rug without ever tapping a toe.

CENTIPEDES

If a centipede is on its last legs, it still may have a long and productive life ahead of it. The word centipede is Latin for "one hundred legs," and in some cases not only is it not an exaggeration, it's an understatement. Centipedes, despite being verminous, venomous, and many-legged, are not insects at all; they're literally in a class by themselves, called the Chilopoda. Chilopods differ from insects in several fundamental ways. First, insects have six legs or fewer; centipedes never have fewer than thirty. Insects have three main body divisions, with legs attached only to the thorax (or central portion); centipedes have a pair of legs on every body segment, from the first one behind the head all the way to the rear. Finally, centipedes have a pair of appendages on their first body segment that functions as a set of poison jaws. The poison jaws are used by centipedes to capture prey; even small animals are sometimes seized in the clawlike jaws and paralyzed. For most centipedes, prey consists mostly of insects, although the larger tropical forms are reported to kill and eat mice, small birds, and geckos. The larger tropical forms are indeed large—one species of *Scolopendra* can exceed ten inches in length.

On the local scene, the most commonly encountered centipede, *Scutigera coleoptrata*, is usually about two inches long when full grown. It's greyish brown with fifteen pairs of long skinny legs (although it moves so fast, they're hard to count). The last two legs point backwards and are longer and skinnier than the other twenty-eight. *S. coleoptrata* is normally found under rocks and stones throughout most of eastern North America, but it's also the species that occasionally sets up res-

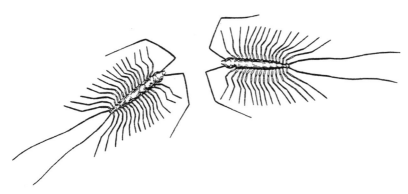

Centipedes: "Bad news, dear . . . Junior needs shoes again."

idence in basements of houses. These centipedes are generally found in the neighborhood of drains and sinks, where they spend their time engaged in the praiseworthy business of catching and consuming cockroaches, clothes moths, and house flies.

Centipedes have a bad reputation largely because of their sinister over-leggy appearance and suspicious habit of scurrying frantically whenever the lights are turned on, as if they were engaged in something to be ashamed of. In point of fact, they're quite benefical in that they kill a number of household pests that do eat paper products, wool, food, and other household possessions. While it's true that centipedes can inflict a painful bite if handled, there's some consolation in the fact that the bite of a centipede, even the foot-long varieties, is seldom if ever fatal to humans. If nothing else, people ought to admire the centipede for orchestrating and coordinating all those legs. They can move a lot faster than a lot of insects with only six legs. The mechanics of the problem have been immortalized in an anonymous verse, perhaps the only poem ever written about centipedes:

A centipede was happy quite,
until a frog in fun
said, "Pray which leg comes after which?"
This raised her mind to such a pitch,
she lay distracted in a ditch
considering how to run.

CLOTHES MOTHS

Coming out of the closet is a very different sort of proposition for insects than it is for people. For the clothes moth, it's an annual occurrence. Each year, when people get around to unpacking their winter woollies for the long cold haul, the clothes moth sees the light of day. Clothes moths and their allies belong to the family Tineidae, a rather large family of rather small moths. The adults, sometimes called millers, as a rule don't feed at all and don't fly very well, either. They measure a mere half-inch across and are generally a nondescript yellow or beige with a glossy sheen. The two most common species, the webbing clothes moth (*Tineola bisseliella*) and the casemaking clothes moth (*Tinea pellionella*), can be distinguished by the presence of dark brown spots on the forewings of the casemaking clothes moth and the absence of said spots from the webbing clothes moth.

Even among insects, the diet of the immature clothes moths is unique. They (and carpet beetles) are about the only organisms in the world that can digest keratin, a rather refractory protein produced by animals.

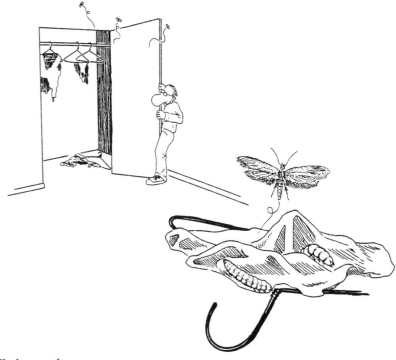

Clothes moths

This ability gives the clothes moth caterpillar carte blanche when it comes to animal products, including wool, hair, fur, feathers, upholstery, milk, fish meal, dead insects, leather, and bristles. Adult females lay eggs one at a time, up to about 150, on any suitable food source, and the caterpillar hatches about five days later. Since the adults dislike light and are easily disturbed, this will all take place, likely as not, in a dark, undisturbed area like a storage closet. The length of larval development varies with conditions and can take anywhere from six weeks to four years. The casemaking clothes moth, *Tinea pellionella*, constructs for itself a thin, papery, silken case which it drags around everywhere it travels and into which it retreats if disturbed. Instead of a case, the webbing clothes moth, *Tineola biselliella*, constructs a stationary tube, saving on silk by incorporating whatever material it's feeding on into the tube. A related species, the carpet moth (*Trichophaga tapetzella*), prefers feeding on carpets, not clothing.

Clothes moths are essentially shy and almost never feed on clothing that is regularly worn. Since they are restricted by physiology to animal protein, they won't touch plant fibers such as linen or cotton or synthetics such as rayon. Despite the fact that it is an animal product, silk is not one of the preferred foods of the clothes moths, perhaps out of deference to its producer (a fellow insect, *Bombyx mori*, the silk moth). Since even clothes moths require a balanced diet, newly hatched caterpillars prefer material that is stained or soiled by food, sweat, and saliva; these materials, as well as the bacteria growing on them, provide the caterpillars with essential vitamins and nutrients.

Control measures for clothes moths involve prevention more than anything else. Clothing to be stored for a long period of time should be dry-cleaned or brushed, shaken, and hung outdoors in bright sunlight. Cedar chests do afford some protection as long as the cedar's fresh and the lid is tight-fitting. Far more effective are "moth balls," naphthalene or paradichlorobenzene (PDB) crystals, interspersed among the clothes. These, however, are repellents and prevent establishment rather than eliminate residents. Thorough vacuuming and dusting help to remove lint and woolen fiber, which may act as reservoirs for future infestations. To control a persistent infestation insecticides must be applied, and it's probably best to get professional help to find out what is safest to use (insofar as you are likely to wear it next to your skin if you treat your garments). A few treatments with the right kind of chemicals will ensure that the caterpillars are "died in the wool," so to speak.

CLUSTER FLIES

In winter, as snow begins to fall and temperatures begin to drop, people tend to spend much more time indoors. Not surprisingly, there are insects that do the same thing. Perhaps the most conspicuous insect that stays housebound for the winter months is the cluster fly, *Pollenia rudis*. The cluster fly looks like a house fly, albeit a bit larger, a bit darker in color, and a bit fuzzier (with short, crinkled golden hairs on the thorax), and it lives in houses like a house fly. Its closest two-winged relatives, however, are bluebottles, blow flies, screwworm flies, and other species in the family Calliphoridae. Cluster flies can be found on at least three continents (North America, Africa, and Europe, for those keeping score), and in North America they range as far south as northern Florida and as far north as Nova Scotia.

Throughout the warm summer months, cluster flies lead an out-doorsy existence, lazily visiting flowers (often wild parsnip and Queen Anne's lace), and sunning themselves on fence posts, tree trunks, and telephone poles. Unlike the majority of their calliphorid relatives, cluster fly maggots do not develop in carrion, manure, or open wounds of cattle and sheep; they've essentially gone underground. Starting in April, females begin laying eggs, two or three at a time, in the soil. When the larvae hatch in four to six days, they penetrate the bodies of passing earthworms, generally through the conveniently prefabri-

Cluster flies

cated male genital opening or thereabouts. The growing larva then goes about the business of consuming the earthworm's entire body contents. Since the atmosphere is anything but fresh inside a worm, cluster fly maggots early during their residence perforate the worm's body wall at the anterior end and position their rear spiracles (or respiratory openings) facing the outside world for a breath of fresh air now and again.

The larval stage of *P. rudis* lasts about two weeks, after which time the larva burrows out of its still-living host to pupate in the soil. Pupation lasts about thirty-nine to forty-five days. Cluster flies have about four generations a year in the U.S.; those emerging from the soil in late September usually hibernate through winter, and the location of choice is inside a house. Cluster flies aren't really intrusive enough to be considered truly pestiferous. They cause greasy stains when they're smashed by irritated homeowners and make loud buzzing sounds when disturbed but barring those habits are mostly only annoying. Their overwintering aggregations can reach huge numbers, however, especially where flies can gain access through unobtrusive, out-of-the-way openings (such as where window ropes pass through their tracks).

Once inside, one place is as good as the next for settling in, and no sort of structure is passed up by cold cluster flies. As Dall wrote in 1882, "People soon learned to look for them (the flies) everywhere; in beds, in pillow slips, under table covers, behind pictures, in wardrobes nestled in bonnets and hats, under the edge of carpets, and in all possible and impossible places. A window casing solidly nailed on will, when removed, show a solid line of them from top to bottom; they are uncanny. They like new houses, but are often found swarming in old unused buildings and go regularly to church, or perhaps only a few good ones abide in sanctuaries; anyway they are there." One imagines that, if some hang out in church, there are undoubtedly others which pass their winter in taverns and saloons. There, they undoubtedly would fit right in with all the other barflies.

COBWEB SPIDERS

Most people who think they live alone probably don't. All it takes is a few weeks of not dusting in corners or basements to reveal that one is usually the unwitting landlord for an assortment of cobweb spiders. Cobwebs are the silken webs of a variety of spiders, most of which are in the family Theridiidae. The majority of theridiids are small spiders that spin webs on plants, in abandoned rodent burrows, or under debris, but a select few—*Theridion tepidariorum* to name one—prefer to set up housekeeping in people's houses (or barnkeeping in people's barns). The theridiids are called the comb-footed spiders because there is a toothed, comblike structure on the last segment of their fourth pair of legs (unlike insects, spiders have eight, not just six, legs). The comb is used to throw a line of silk out after its prey. Like all spiders, cobweb spiders are carnivorous, and the cobweb, which from a human perspective seems to serve no purpose other than to collect dust, is actually a cleverly designed snare for unsuspecting insects.

The theridiid web is far more intricate than it appears at first glance. There is usually a central dense mat or maze, used by the spider as a retreat or resting place. A series of long guy lines, held taut by tiny silk studs, anchors the whole thing in place. When an insect stumbles into one of the sticky guy lines, it breaks and coils, lifting the body up

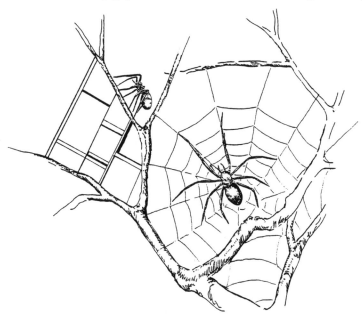

Cobweb spiders: "He'll never get anywhere with that modern web stuff."

into the air. The spider, sensing the disturbance in the web, rushes out, backs up, and using the comb on its hind legs, draws out a thick heavy strand of silk to tie down the insect. It then turns around and bites the hapless insect with its fanglike chelicerae, or jaws. In the process, the spider injects a paralyzing venom. The victim soon ceases struggling, and the spider transports it back to the maze, to suck it dry at its leisure. Only very rarely does an insect escape after it has been snared—among other things, spider silk has a tensile strength second only to fused quartz fibers, making it difficult (to say the least) for a captive insect to snap. Silk is produced by three to four pairs of organs called spinnerets projecting from the base of the abdomen. The silk is so strong, and the theridiids sufficiently gifted with engineering prowess, that the common cobweb spider is capable of subduing and lifting small snakes, mice, and other major-league prey items. By far and away, however, small insects suffice for dinner.

The inclination of theridiids to rush out, snare, and eat anything that creates vibration in the web makes mating a precarious process. Among other things, the male is minute in size. *Theridion tarrenfosorum* females, for example, are almost one-third of an inch in length, while the males scarcely measure one-eighteenth of an inch. Males approach a female from the nether portions of the web in order to ascertain her mood. Those that are incautious are occasionally eaten. Spider mating in general is an awkward process anyway. The males have an intromittent organ in the end of the leglike appendages (palps) next to its head. These palps, resembling a hypodermic needle in overall structure, have no connection to the sperm-producing organs, so males must spin a little "sperm web," deposit sperm on it, and then suck it into their needlelike palps in anticipation of injecting it into some willing (and replete) female. After laying eggs, the female spins a silken sheet and shapes it into a sac. After the spiderlings hatch, having molted once in the egg sac, they spin a thin strand of silk and in paratrooper fashion drift away on the next breeze.

While silk is central to the life of the spider, it's more of an annoyance to the average homeowner. The sticky silk, designed to catch small insects, also catches dust, and cobwebs soon grow to be unsightly messes to be dispensed with at one's earliest convenience. Such was not always the case, however. Cobwebs were once in great demand in Europe as bandages to stanch blood flow. In the nineteenth century, members of the Burgman family of Innsbruck were so taken with the texture of cobwebs they used several layers of clean web as a canvas for painting. Some of these fine paintings are still on exhibit in American collections, although all are undoubtedly unsigned by the spiders.

EARWIGS

Much of the bad press received by earwigs they bring upon themselves. First of all, in terms of appearances they're not exactly as cute as bunnies. Earwigs, members of the order Dermaptera, are brownish-to-blackish, medium-to-large insects generally sporting at the tip of the abdomen an evil-looking pair of enormously enlarged forceps-like pincers. Appearances in this case are not deceiving—they can give you a nasty pinch if they consider themselves mishandled. To add insult to injury, many species of earwigs possess glands on their third and fourth abdominal segments from which they can secrete, or even shoot over some distance, vile-smelling repugnatorial secretions, the putrid odor of which persists for what seems like forever.

In reality, however, there is much about the earwig to admire. There's no truth to the rumor that earwigs crawl into people's ears at night to wreak all sorts of unimaginable havoc. The name "earwig" probably refers to the fact that the hind wings, folded and tucked up underneath

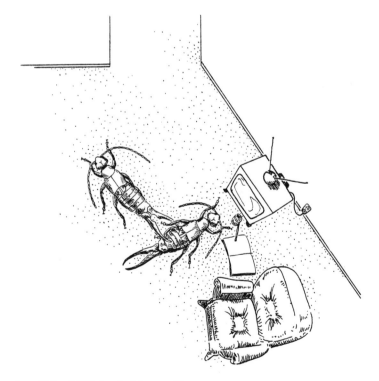

Earwigs: "Mom! Billy's pinching me!"

the short, truncated leathery forewings, are vaguely earlike in shape—hence, "earwi(n)g." The wings are of little use for flying, and, as a consequence, earwigs can can only get airborne by lauching off a high place. They therefore seldom travel long distances under their own power.

This is not to say, however, that they don't get around. Of about nineteen common North American species, at least seven immigrated from Europe, usually as stowaways in ship cargo. America's most common earwig, *Forficula auricularia*, was first reported in Seattle in 1907. It now can be found in and around the gardens, basements, and garbage piles of homes across the country.

F. auricularia is approximately three-fourths of an inch long and possesses the reduced wings and enlarged forceps typical of earwigs in the suborder Forficulina. Adults emerge from underground nests in July. These adults mate, and females lay a clutch of about thirty eggs underground. Unlike most insect mothers, female earwigs are doting parents; in many species, mothers devotedly guard their eggs and developing offspring from potential enemies until they're ready to leave the nest. Development is complete in approximately two months. The second generation lays eggs underground in January or February, and the eggs hibernate until they hatch around April.

Earwigs are not well loved by most people because as garden pests they can cause extensive damage to flowers and vegetables. They also have sufficiently bad manners to enter houses and other human habitations without an invitation. They appear so frequently in corn processing plants that they're sometimes called "starch bugs." However, in their defense, it can be said that they may be working on the side of the homeowner, unbeknownst to him or her. Earwigs as a rule are tremendously omnivorous, and in addition to vegetable material they are not averse to eating other insects. They ravenously devour all sorts of arthropods, including aphids, gnats, grubs, and even fleas. So if you should run across an earwig in your basement or garden, remember that you may require their insect-eating services someday—in a "pinch."

HOUSE-DUST MITES

If I told you that *Dermatophagoides pteronyssinus* bit the dust, you'd probably think that some dire fate had befallen it. Actually, when *Dermatophagoides* bites the dust, it's having breakfast. *Dermatophagoides* mites are the chief residents of the house-dust ecosystem and are among the more common arthropod inhabitants of human habitations. In one study in Holland, *Dermatophagoides* were found in every one of 150 houses surveyed, and they probably occur in every house in the world that needs dusting. That they're not as widely recognized or despised as cockroaches, silverfish, or other domestic insects is no doubt due to the fact that they're essentially microscopic in size, only about one-hundredth of an inch in diameter. Overlooking them is thus eminently forgivable. *Dermatophagoides* and its kin are not insects but rather are mites, arachnids in the order Acari. Like spiders and other arachnids, they have eight legs instead of six at maturity. Their closest relatives include such acarine delights as the ear mites of cats and dogs and the scab mites of sheep.

House-dust mites: "From dust ye came and to dust ye shall return."

The life of *Dermatophagoides* is literally dust to dust: they're born to dust, live in dust, and die in dust. All life stages (eggs, six-legged larva, eight-legged protonymph, tritonymph, and adult) can be found in house dust. At times population densities exceed five hundred mites per gram of dust, or roughly fourteen thousand mites per ounce of dust. Among all the constituents of dust, including skin scales, textile fibers, pollen, fungi, algae, food crumbs, and the like, *Dermatophagoides* is particularly fond of human skin scales—dandruff in particular, but shed skin from other body parts is equally acceptable. Unadulterated dandruff, with its fat content of 10 percent, is actually toxic to the house-dust mite; only partially defatted dandruff will do. Another occupant of house dust, *Aspergillus penicilloides*, a fungus, obliges the mites by predigesting the dandruff scales and lowering the fat content so that *Dermatophagoides* can dine. While the mite rarely occurs without the fungus, the fungus can occur without the mite, and in fact at high humidites (80 percent and above) grows so abundantly that the mites aren't left with enough dust to support themselves. Whoever said "dry as dust" wasn't kidding, and probably the greatest challenge to the house-dust mites is to find an environment with a humidity high enough to support life but low enough to slow down the growth of *Aspergillus*. Optimal humidity is about 71 percent, the same relative humidity that most people find comfortable, so *Dermatophagoides* mites are amiable roommates.

Infestation with house dust-mites isn't quite as innocuous as it may appear. *Dermatophagoides* species are the chief agents responsible for "house-dust" allergies, and inhaling them is thought to be a factor in some asthma attacks. Moreover, at times *Dermatophagoides* doesn't wait for the skin scales to fall where they may but rather invades the skin to cause allergic dermatitis. While they themselves can't penetrate the skin, they can set up residence at scabs, ulcers, fungus infections, and other skin eruptions and there add insult to the original injury. So there's more incentive for dusting the house than simply aesthetics. There might even be incentive to buy a good strong dandruff shampoo.

TERMITES

Considering how familiar an insect the termite is, there's an amazing amount of misinformation about it in circulation. First of all, termites are called white ants, yet every spring, the winged termites that are commonly seen aggregating around windows and around the bases of outside walls are black, not white. Furthermore, they're not ants; despite certain superficial resemblances, ants and termites are really quite different and are easily distinguished. Ants have wings with only a few veins, with the hind wing much smaller than the front wing. In contrast, termite wings have many veins and are the same size (hence the scientific name for the order that termites belong to, the Isoptera, meaning "wings of equal size"). Moreover, termite wings are long, often twice as long as the body; ant wings are rarely if ever that long. Ant antennae are also distinctive, with a jaunty elbow-shaped bend; termite antennae are conservatively straight. Finally, ants have a characteristic hourglass shape, the result of what's known generally (if not entomologically) as a "wasp waist." Termites have more of a "chemise" waist, not noticeably indented at all.

Termites are similar to ants in that they're social insects, but even the structure of their society is very different. For example, termite colonies are unique in having not just a "queen" but a "king" as well. These functionally fertile males and females (variously called primary reproductive imagoes or alates) fly in swarms during the spring, usually after a warm rain. Individuals separate into pairs and land on the ground, at which point their wings fall off along a prearranged suture line. The king and queen together excavate a cell in soil or wood and then mate. King and queen termites pair off for life, mating repeatedly for the duration of their lives, in the manner of English royalty before

Termites

Henry VIII. Bee queens, in contrast, mate with many males, all of whom die in the act since their genitalia are ripped off by the process. After mating, the termite queen turns her attention to egglaying; in some species, this attention is so fanatic that she loses the ability to walk. The eggs laid by the queen develop into four classes or castes of termites: (1) additional primary reproductives at swarming time, (2) secondary reproductives (pale insects with wing buds instead of wings that can lay eggs to supplement the queen's, although they are incapable of producing primary reproductives), (3) sterile workers (the grayish "white ants" that form the bulk of the colony and take care of quotidian chores such as nymph rearing, cleaning, and climate control), and (4) sterile soldiers. Workerlike in appearance, the soldiers have enormously elongated jaws, or mandibles, and a head packed with muscles to help them take care of the defense of the colony. In some tropical termites, the jaws are replaced by a conelike structure through which termites can shoot a caustic sticky substance at intruders. Nymphs or immatures are cared for generally by the adult workers and reach maturity in three to four months or more.

Another popular misconception about termites is that they eat wood. Granted, they do a lot of damage to wood, but the majority of termites are no more able to digest wood than you or I. In their guts, termites house a variety of microorganisms that produce an enzyme called cellulase. This enzyme breaks down the major structural component of wood, cellulose, into bite-size sugar molecules that the termite can digest. In exchange for the enzyme the microbes obtain a safe, warm, amenable place to live—the termite gut. The microbes can be acquired only by ingestion, and parents pass the microbes directly to the offspring by a process called anal feeding (which is exactly what it sounds like). This complete dependence of termites upon their microorganism-partners has been suggested as the reason behind the termites' social habits.

Probably 95 percent of the termite problems in the continental U.S. are caused by *Reticulitermes flavipes*, one of the subterranean termites. Subterranean termites require moisture and contact with the ground. The other main class of termites, drywood termites, do not. Moreover, subterranean termites "eat" wood along the length of the grain, not across the grain (as the drywood termites do). This physiological requirement for moisture accounts for the dirt tunnels that *R. flavipes* constructs for itself. These tunnels allow the termites to move around freely in a climate-controlled environment.

Breaking open these tunnels or tubes is the first step in controlling termites. Basically, there are four steps involved in controlling termites. Most important is prevention, creating moisture and feeding conditions unfavorable to termites. Draining standing water, improving ventilation under buildings, and removing stray wood stumps, stakes, and scraps are all important. Second and third are soil and foundation treatment. This is generally done by professionals and involves application of persistent insecticides. Finally, wood can be treated with a protective chemical such as pentachlorophenol.

Considering how destructive termites are of human habitations, it's somewhat surprising that termites can themselves be regarded as the master architects of the insect world. Termite nests in the tropics, which can house upwards of a million individuals, may reach fifteen to thirty feet in height and diameter and may contain hundreds of galleries, nurseries, and chambers. Termites have adopted not only traditional architectural innovations like the column and the arch, but also very modern engineering techniques as well. Termite mounds of various species make use of waterproofing, insulation, passive solar heating, and even air-conditioning. More remarkable is the fact that, unlike with humans, all of this work goes on without the need for mortgage financing or home improvement loans.

Chapter 3

Garden-variety Types

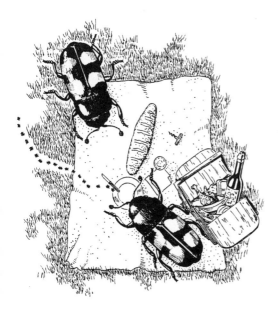

Bean leaf beetles

Black swallowtails

Junebugs

Mexican bean beetles

Picnic beetles

Plant-feeding stink bugs

Spittlebugs

Squash bugs

Tobacco hornworms

BEAN LEAF BEETLES

If you think you're seeing spots before your eyes in the early fall, you're probably right—bean leaf beetles are out in force. Bean leaf beetles, known to their friends and scientific associates as *Cerotoma trifurcata*, range in size from a fifth to a fourth of an inch in length and are variously red, green, tan, yellow, or brown. They are also variously marked, frequently by a black border around the margin of the elytra, or wing covers, and by two pairs of spots along the inner wing margin. In some individuals the spots are missing altogether, and in others the spots run together and are perhaps better described as splotches rather than spots.

While the appearance of the bean leaf beetle may vary, its habits are a lot more predictable. Bean leaf beetles, as one might guess, can almost invariably be found on bean leaves—if not on bean leaves, then bean blossoms, pods, or stems. To the bean leaf beetle, "bean" means just about any plant in the family Leguminosae, and *Cerotoma trifurcata* has made its presence known on soybeans, green beans, cowpeas,

Bean leaf beetles

peanuts, bush clover, beggartick, tick trefoil, and a host of other legumes.

In the spring, the beetles emerge from their winter quarters and set about the business of riddling leaves, generally from the underside. After a few weeks, the females begin to lay their orange-colored eggs in clusters of one or two dozen in the soil at the base of the plant. The eggs hatch and the larvae eat all the remaining parts of bean plants—the nodules, the roots, and the stem below the ground. The larvae are white segmented grubs, brown at both ends with three tiny pairs of legs toward the head end of the body. In three to six weeks, the mature larva constructs a chamber in the soil and pupates; an adult emerges after about a week, and the beetle resumes feeding on bean leaves until the weather turns cold and the leaves begin to yellow. At that point, beetles leave the field and begin frantically looking for a place to spend the winter in comparative comfort—in ditches, banks, wooded areas, fencerows, and the like.

Bean leaf beetles display a rather distinctive behavior that is characteristic of many beetles in their family, the Chrysomelidae or leaf beetles. If the plant they're sitting on is disturbed, they respond by curling up and "dying," since at least pretending to die is a very effective defense against their enemies, many of which locate their prey by watching for movement. This defense, widespread among the Chrysomelidae, has somewhat unjustly been incorporated into the English language as "playing possum." On the basis of numbers alone, however, given that upwards of twenty thousand chrysomelids inhabit the world, it should more rightly be called "playing beetle."

BLACK SWALLOWTAILS

Parents whose children are reluctant to eat their vegetables, be advised that there can be too much of a good thing; parents of black swallowtail caterpillars can't get them to eat anything but a few select vegetables. The immature black swallowtail (mostly known as the celeryworm, carrotworm, or parsleyworm) would actually starve rather than eat any plant outside the family Umbelliferae. No immediate danger of that exists, however, insofar as the Umbelliferae is rather large as plant families go. More familiar members include carrot, parsley, parsnip, celery, anise, dill, fennel, lovage, chervil, cumin, coriander, caraway, and a host of other spice and vegetable plants.

The immature caterpillars are black, with a white saddle mark mid-length. For want of a better description, at this point in their lives they most closely resemble a bird dropping. This is probably not a coincidence, since birds are major predators of caterpillars and no doubt, at casual glance, would not pursue a bird dropping as a potential meal. The ruse works, one supposes, as long as the caterpillar remains motionless, since bird droppings don't generally locomote (and even birds aren't that stupid).

After three molts, the black swallowtail changes color and becomes green with black stripes alternating with yellow markings. It passes through one more molt in this outfit and can reach lengths approaching

Black swallowtail: "Why parsnips, Lord?"

49

two inches in the process. The reason for the color change in mid-life hasn't really been resolved yet, but it may be that the bird dropping gimmick wears thin on a two-inch caterpillar, unless it happens to be feeding in an area populated by very large birds. The green-and-black-striped mature larvae, however, are not entirely without defenses. The caterpillars have a Y-shaped organ tucked behind their head which can be popped out and directed menacingly at intruders. Adding to the effect is the fact that this organ produces several volatile and obnoxious chemicals, including isobutyric acid, the agent responsible for the aroma of rancid butter.

At the fifth molt, the larvae turns into a pupa that generally is green in summer and brown in fall, to match the surrounding vegetation. In summer the butterfly emerges in about ten to fourteen days; in the fall the pupa goes dormant for the winter, and the butterfly doesn't emerge until the following spring. The adult butterfly is a pleasant surprise, considering it began life resembling a bird dropping. With a wingspan of three inches or more, the black swallowtail and its relatives are among the largest butterflies in North America. It owes its name "swallow-tail" to the two tail-like projections of the hind wing. The butterfly is black, with two rows of yellow spots near the front wing margin; females generally have a blue patch and both sexes have a row of orange spots and an inner row of yellow spots on the hind wing.

The sex life of the black swallowtail has recently attracted scientific interest. In spring, males emerge two to three days earlier than females and cruise the neighborhood looking for a territory. More often than not, prime territory is on top of a hill. The first male to arrive immediately sets out to defend his hilltop from intruding male black swallowtails, and the resident male ferociously attacks and drives away any other black swallowtail that presumes to interlope. Resident males are so obsessed that they have been known to pursue red-winged blackbirds and black T-shirts as well. When the females finally emerge, they head for the hills, where the victorious male awaits to greet them and mate. As far as is known, males that fail to establish a territory also fail to mate. One supposes that the females simply regard them as "over the hill."

JUNEBUGS

A more or less timely topic every spring is the Junebug, sometimes called the May beetle. Junebugs are really beetles, equipped as they are with the two hardened wing covers, or elytra, characteristic of the order Coleoptera, the beetles. June beetles, then, are the brown to brownish black amazingly clumsy beetles about three-fourths of an inch long that persistently attack your porch light at night. If you wonder why you never see them in the daylight, it's because at dawn they leave the porch and bury themselves in the soil, where they stay until the next evening, waiting for porch lights to go on again.

When not careening into screen doors, adult Junebugs engage in a number of other more productive activities. The Junebugs, which comprise about two hundred species in the genus *Phyllophaga*, belong to a group of beetles collectively called leaf chafers, since the adults feed primarily on the leaves of trees. The very name of the genus, *Phyllophaga*, in Greek literally means "leaf eater." In the case of the Junebugs, the leaves in question include oak, hickory, elm, ash, birch, maple, willow, walnut, pine, and persimmon, to name a few.

Junebug larva

When not devouring leaves, females are busy ovipositing: each female lays up to two hundred eggs in the soil, preferably in places like lawns or permanent pastures. The eggs hatch in two to three weeks, and the larvae of the June beetles, better known as white grubs, begin to feed on the roots of just about anything they encounter. They're particularly destructive to corn, pasture grass, and nursery stock. The fat white C-shaped larvae can be distinguished from a host of other fat white C-shaped soil-dwelling grubs by the double row of tiny spines on the underside of the last body segment. In this way the native June beetles can be told from such imported menaces as the Oriental beetle, the Japanese beetle, and the European chafer, if anyone is inclined to inspect the undersides of grub abdomens.

The length of a complete life cycle varies with the species of Junebug, ranging from one to four years; quite common is a three-year cycle. After their first summer the grubs burrow deep into the soil to hibernate; they continue to feed in their second summer and after a second hibernation feed in the third year until June, at which point they pupate in chambers underground. Although the adults develop in a few weeks, they don't emerge from the soil until the following spring, from May through June, depending on the locality. Hence the calendrical confusion regarding the common name.

The June beetle is occasionally a pest of corn and wheat, and a general annoyance in the home garden. Plowing or tilling between mid-July and mid-August will crush many of the pupae and adults and turn up larvae for crows, blackbirds, and other birds to gobble down. Not too long ago the method of choice for clearing grubs was to run hogs through the field in summer and fall. The hogs would root out and eat the grubs, eliminating the insect problem and getting fat at the same time. It sort of adds a new dimension to the old expression, "Let's rustle up some grub."

MEXICAN BEAN BEETLES

Every family has a black sheep—some reluctantly acknowledged relation with a checkered past or dubious present. For the beetle family Coccinellidae, the ladybugs of nursery-rhyme fame, the odd ones out are species in the genus *Epilachna*. The vast majority of ladybugs, or more correctly, ladybird beetles, are predaceous, feeding voraciously on aphids, scales, and other soft-bodied plagues of humanity. Among humans they are universally regarded as insect good guys; coccinellids have been instrumental in biological control programs since the Vedalia beetle, *Rodolia cardinalis*, saved the citrus industry from cottony-cushion scale in 1888.

But in *Epilachna*, the same single-minded voracity is lamentably directed toward plant foliage. Like its relations in the family Coccinellidae, *Epilachna varivestis*, the Mexican bean beetle, is round, strongly convex, and conspicuously spotted. Copper red with sixteen black spots arranged in three rows along its back and with the three-segmented legs typical for the family, the Mexican bean beetle is anything but unique as ladybird beetles go. Its habits, however, set it apart—instead of masticating aphids and scale insects, *E. varivestis* chews off fragments of bean leaves, wads them into a pulp, and then sucks out the

Mexican bean beetles: "Hola, Paco, como esta?" "Muy 'bean!'"

juice. Most frequently, *E. varivestis* attacks bush and pole beans (*Phaseolus vulgaris*) and lima beans (*P. limensis*), but beggarweeds, cowpea, alfalfa, hyacinth bean, and soybean serve equally well. Azuki bean, clover, and kudzu suffice in a pinch, and in desperate times the Mexican bean beetle can abandon legumes altogether for squash, eggplant, and even okra.

About the only reliable characteristic separating *Epilachna* from its less obnoxious relatives is the distinctive appearance of the larvae. Mexican bean beetle larvae are sluggish yellow grubs covered with forked spines. At maturity, they measure about one-third of an inch in length. As youngsters especially, the grubs scrape leaf tissue from only one side of the leaf, leaving as an unmistakable signature foliar "windows" that dry out and crack. Older larvae and adults with larger mouthparts eat everything but the veins, leaving leaf skeletons in their wake.

Adult Mexican bean beetles spend the winter under litter in woodlands, generally not too far from their home bean fields. Should woodlands be in short supply, however, fenceposts, stone piles, woodpiles, and garden refuse provide adequate shelter. Then when beans first appear in early spring, the overwintering beetles are there to eat them. After a week to ten days, females lay yellow eggs in batches of 40 to 60 on the undersides of leaves—one female can lay anywhere from 400 to 1,600 eggs. In spring, eggs hatch in ten to fourteen days; eggs laid later in the season in the height of summer heat hatch in half the time.

Larvae undergo four molts and reach the pupal stage in slightly over a month's time, depending on local conditions. Unlike many leaf-feeding beetles, *E. varivestis* and the rest of the lady beetles do not pupate in a concealed place, as in the soil or under litter—the pupa remains defiantly anchored to the leaf, with the last larval skin still attached at the base. After one to three weeks, adults emerge from the pupal skin, and, in a little over a week, females are back laying eggs if good weather prevails. In the South, Mexican bean beetles run through three or four generations each season; in the North, time permits only a single generation.

The Mexican bean beetle, though, is not the only (exo)skeleton in the coccinellid closet. *Epilachna borealis*, the squash beetle, shares the bean beetles' predilection for plants, although it is partial to the leaves of pumpkin, squash, and other members of the family Cucurbitaceae. It is also one-fourth to one-third of an inch in length, yellowish, and convex. It is also bedecked with seven large black spots on each wing cover but has four black spots on the thorax that are absent from *E. varivestis*. By the same token, *E. varivestis* is not the only spotted beetle that attacks beans. The bean leaf beetle, *Cerotoma trifurcata*, though in the family Chrysomelidae, can match the Mexican bean beetle spot for spot. Similar in shape, it is slightly less convex and considerably smaller (topping off at about one-fifth of an inch).

Mexican bean beetles were first described in 1850 from specimens collected in Mexico. Their introduction into the United States, first documented in 1853 in Colorado, might well have been a consequence of the Mexican War, when large amounts of hay were moved across borders to fill the needs of hungry cavalry horses. The beetles moved east into Alabama in 1918, again in a shipment of hay, and from there they spread extensively in all four compass directions. Now Mexican bean beetles are found in virtually every state east of the Mississippi, in much of the Southwest, throughout the Midwest, and as far north as Idaho.

Recent host switches throughout their range to soybeans have escalated the status of Mexican bean beetle from occasional annoyance to the home gardener to legitimate economic pest. According to the terms of the treaty that terminated the Mexican War, the Treaty of Guadalupe Hidalgo of 1848, the United States paid Mexico $15 million for all of the territory north of the Rio Grande and Gila River from the Colorado to the Pacific. Without the benefit of a treaty, however, the Mexican bean beetle succeeded in adding most of North America to its range as a result of the Mexican War—so it looks as if ladybugs could teach the politicians a thing or two about Manifest Destiny.

PICNIC BEETLES

This short subject is indeed a short subject. Picnic or sap beetles seldom exceed one-fourth of an inch in length. The family name, Nitidulidae, is from the Latin word for "bright" or "shining" and refers to the glossy appearance of these beetles, which come in a variety of bright shiny colors. Nitidulids have an even greater fondness for picnics than do ants and regularly appear at backyard barbecues, picnics, roadside fruit stands, and garbage dumps. *Glischrochilus sanguinolentus*, a remarkably weighty name for a beetle one-tenth to one-fifth of an inch long, is bright red with a black spot on each wing cover; *Glischrochilus quadrisignatus*, a close relative, is reddish with a pair of yellow spots on each wing cover. Collectively, these and other species in the genus make up the picnic beetles, which have the annoying habit of crawling in and around the tips and cracks of sweet corn and in cracks and injuries in tomatoes, apples, and other bruised fruits and vegetables. They are, on occasion, pests in gardens, particularly on ripe fruit that

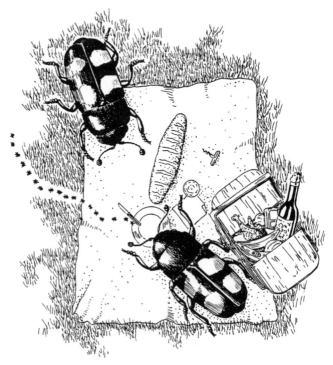

Picnic beetles: "Ants! It never fails."

has fallen to the ground, but they've been known to infest undamaged apples on the tree as well. Species in the genus *Carpophilus* are called sap beetles. *Carpophilus dimidiatis*, the corn sap beetle, is blackish red to brownish yellow, with an orange spot on each wing cover. It and its relatives can be found in corn, in the field or on the table, crawling around tips and cracks; they generally first appear in the field when corn begins to tassel.

Nitidulids are short subjects not just by virtue of their size but because virtually nothing is known about their biology. It's not really clear what they eat exactly. Some appear to feed on the bacteria and fungi associated with rotting or damaged fruits and vegetables; some eat fungi outright; others eat the insects that damaged the fruit in the first place; and at least one species is found almost exclusively on the bones and skin of dead animals (and what *it's* doing is anybody's guess). As for the life cycle, nobody has really figured out the specifics, just that there's one generation per year in the North, they pupate underground, and spend the winter in hibernation. Adults become active in the spring just about when people begin to break out the picnic baskets.

One additional overwhelmingly obvious characteristic of many members of the family is a weakness for alcohol. They favor fruit that has fermented and can easily be lured into an area and trapped with fermenting malt, fruit, or molasses as bait. This seems to fit in well with their predilection for picnics; after all, who hasn't been lured to a picnic or barbecue with the promise of a couple of nice cold beers?

PLANT-FEEDING STINK BUGS

Most families have a few members given over to various forms of eccentricity, and the Pentatomidae, or stink bug family, is no exception. The Pentatomidae is a large family of true bugs—insects equipped with a long beak used for sucking up fluids. The name "Pentatomidae" refers to their five-segmented (hence *penta*) antennae. But while two of the three subfamilies prefer to sink their beaks into insect prey, members of the subfamily Pentatominae are sixties-style vegetarians.

The plant-feeding pentatomids vary in their approach to vegetables. The harlequin bug, *Murgantia histrionica*, for example, feeds exclusively on plants in the crucifer or cabbage family, favoring cabbage, broccoli, turnip, kale, and horseradish. These flashy shield-shaped, black-and-red bugs use their sucking mouthparts to draw out plant chlorophyll and vital fluids, leaving behind yellow-and-white blotches on the foliage, which then looks for all the world as though it's been scorched by fire (the harlequin bug is in fact called, on occasion, the firebug). The whitish barrel-shaped eggs of the harlequin bug are equipped with black "stays," and even a black bunghole, and are laid in double rows of a dozen or so on the undersides of leaves. Eggs hatch in a week to a month, and after five instars (in about two months) development is

Plant-feeding stink bugs: "For me? Deodorant?"

complete. The adult insect is a major pest of cole crops in the North (that's *cole* as in coleslaw—cabbage, brussels sprouts, and the like), although it rarely causes damage above the fortieth parallel. The story is that Union troops unleashed it on the rebels during the Civil War, ostensibly to undermine morale by sabotaging the turnip green supply. The vegetarian gourmand among the Pentatomidae is *Nezara viridula*, the southern green stink bug, which will never turn its nose up (or its beak down) at any field crop. It avidly eats tomatoes, peas, turnips, alfalfa, cotton, soybeans, pecans, peaches, apples, cherries, eggplant, okra, cabbage, and corn, among other things.

The southern green stink bug is aided and abetted in its assault on produce by the green stink bug, or soldier bug (*Acrosternum hilare*), a closely related species with similarly undiscriminating tastes. The two species puncture the skins of fruit with their beaks to cause a type of damage called catfacing or dimpling; their feeding causes seeds to shrivel, and in pecans it brings on a bitter flavor. In general the green stink bugs overwinter as adults in sheltered areas such as roadsides, fencerows, or ditches; their barrel-shaped eggs are laid in spring, and the nymphs hatch out in about a week. Like the harlequin bugs, the green stink bugs go through five instars and reach adulthood in six to eight weeks. Depending on the latitude, stink bugs can go through as many as three generations a year, often sucking well into September.

Although the pentatomids may vary in their diet, there is one family resemblance that crosses culinary lines. Irrespective of their taste in food, most stink bugs do indeed stink. Most members of the family have a pair of scent glands on the third segment of the thorax; these glands are repositories of all manner of noxious substances, released whenever the stink bug feels threatened (in keeping with the philosophy that the best defense is a good offense). The southern green stink bug, for example, pumps out in its glandular secretion undecane, 4-oxo-trans-2-octenol, dodecane, tridecane, 2-propenal, 2-butenal, 4-oxo-trans-2-hexenal, trans-2-hexenal, 3-hexanone, 2-nonanone, trans-2-octenal acetate, trans-2-decenyl acetate, and at least a half-dozen additional odor components—an accomplishment in toxic waste production that could put the U.S. chemical industry to shame.

SPITTLEBUGS

Every spring, in meadows and fields throughout eastern North America, *Philaenus spumarius* busily engages in activities that in New York City subways people regularly get ticketed for doing: *P. spumarius*, the meadow spittlebug, spends its formative days producing a frothy white secretion that could easily pass for spit. In the fall, females lay egg masses of one to thirty on the stems and stubble of any of over four hundred species of plants. The next spring, when the weather warms up, the eggs hatch, and the tiny nymphs immediately insert their syringelike mouthparts into the circulatory system of the plant and pump the vascular fluids of the plant through their bodies. At their posterior end, the copious volumes of liquid waste are discharged with air and viscous glandular secretions to produce a bubbly mass that gradually accumulates around the insects. One reason there is so much liquid to discharge is that, when the spittlebug taps the plant, it hits not the nutrient-rich phloem sap, which is 10 to 25% sugar and up to .4% amino acids, but instead hits xylem sap, a meager .005% sugar and .0002 to .08% amino acids. Needless to say, *Philaenus spumarius* has

Spittlebug: "No spitting on the sidewalk, bub!"

to pump fluids furiously in order to obtain enough nutrients to survive and runs through approximately 150 to 300 times its own weight every twenty-four hours.

Since their diet is so low in nutrients, spittlebugs are, as far as insects go, relatively slow-growing. The nymphs spend at least a week in each of five instars, pumping frantically up until the last molt. The spittle mass is actually more than a waste product. It maintains a high humidity around the soft-bodied delicate nymphs, and it discourages predators, who would have to wade through an appetizer of sticky gobs of spittle before dining. Not surprisingly, the meadow spittlebug has no known natural enemies.

The habit of feeding on xylem sap is bad news for plants, since it is through the xylem that nutrients are brought up from the roots and delivered to the plant's photosynthesizing parts. Forage yields, for one thing (such as alfalfa yields), can be reduced by as much as 33 to 90% in a major infestation. Stunting and wilting can also occur, and, to add insult to injury, there can be problems curing harvested hay because of the additional moisture of spittle masses.

Adult spittlebugs aren't nearly as conspicuous as their offspring. They're a quarter- to a half-inch long and variously brown, grey, or black, with or without spots, and/or striped. They are sometimes called froghoppers because their bluntly angular head, with prominent eyes on either side, is vaguely reminiscent of a frog—although you couldn't really say it's a "spitting image."

SQUASH BUGS

In November the frost may not be the only thing on the pumpkin—
or under it. *Anasa tristis*, the squash bug, has a distinct tendency to
spend the winter in and around the same sort of places it spends spring
and summer—in, on, and around and under almost any plant in the
family Cucurbitaceae, including squash, canteloupe, cucumber, water-
melon, and pumpkins. Adult squash bugs are blackish brown, rather
rectangular insects about one-half inch in length and covered with a
dense blanket of fine black hairs. They feed resolutely and single-
mindedly on leaves, stems, and fruits of all sorts of squashes and
melons, using their long piercing mouthparts to suck sap and liquid
material from stems, leaves, and fruits indiscriminately. Even after a
heavy frost, squash bug adults continue to feed on the sun-facing sides
of fallen fruit. When temperatures get cold enough, they eventually
abandon feeding and fly off in search of a hiding place to pass the
winter—under dead vines, leaves, boards, or stones or even inside barns,
sheds, garages, or houses.

As insects go, they're rather late to get started in the spring. Having
spent the entire winter celibate, the squash bug's first order of business
is to find a mate. Once that's over with, the females deposit glossy
brown elliptical eggs in groups of one to forty on the undersides of
squash leaves in the spaces between the veins. After a week or so, the

Squash bugs: "Then, when the pumpkin is taken inside, we get into the candy
and eat it all."

eggs hatch and the tiny immature squash bugs provide quite a contrast with their sedately brown parents. Their head, thorax, legs, and antennae are bright red and their abdomen is garish green. The young nymphs are very gregarious, spending their time sucking squash sap in the vicinity of their siblings. As they develop, they undergo four molts and end up greyish white with black legs before their final molt into adulthood.

Squash bugs manage to make themselves unpleasant in a number of ways. First of all, like many bugs in the family Coreidae, all life stages of *Anasa tristis* are equipped with glands near the middle of their abdomen. If the bugs are disturbed, they release a foul-smelling substance containing various sorts of pungent volatile aldehydes. Moreover, as they feed over the course of the summer on squash, melons, and other cucurbits, they sap the vital fluids out of the growing plant, leaving drooping, blackened, and, eventually, crisp dead leaves in their wake.

They're not entirely without admirers, however. *Trichopoda pennipes*, a tachinid fly, lays eggs on the body of adult squash bugs. When the eggs of the fly hatch, the maggot chews its way through the eggshell and then through the body of the bug and sets up residence inside the respiratory system of the hapless host. If the fly completes development before cold weather sets in, it crawls out through or near the anus of the bug to pupate outside in the soil; in late fall, the fly larva pupates inside the bug body and spends the winter in hibernation just like the bug.

One of the best ways to control squash bugs is to collect and kill the bugs hanging out around former gardens and then destroy vines, fruits, and any other vestiges of squash life forms before the winter. Yet another way is to plant resistant varieties; although the squash bug likes squashes in general, it has definite preferences among them. Among the less preferred, and therefore more resistant, forms are butternut, royal acorn, pink banana, and black zucchini. This is yet another reason the squash bug is less than well loved—it doesn't even have the decency to cut down on zucchini production.

TOBACCO HORNWORMS

That old line, "Too much tobacco stunts your growth," certainly can't be said to apply to *Manduca sexta*, the tobacco hornworm. The tobacco hornworm doesn't smoke it, though—it *chews* tobacco non-stop, from the moment it crawls out of its eggshell to the time it crawls into the ground to pupate. It can reach four to five inches in length during the process and, not inappropriately, in its final larval instar is more or less cigar-shaped; it's generally light or dark green in color (although black forms crop up from time to time); and it's characterized by a black horn at the end where the filter tip should go.

About the only thing you might mistake for a tobacco hornworm is its close relative the tomato hornworm, *Manduca quinquemaculata*. Both feed on plants in the family Solanaceae (to which not only tomatoes and tobacco belong, but also eggplant, potatoes, peppers, and a few other comestibles), and both are large green caterpillars. However, the larvae can be readily distinguished in several ways: while tobacco hornworms have seven oblique white stripes along the length of their body, the tomato hornworms have eight; moreover, the stripes on a tomato hornworm hook backward on the dorsal (or upper) side to form a V or L shape. In addition, while the horn of a tobacco hornworm is black (like cured tobacco), the horn of a tomato hornworm is red (like a ripe tomato).

It's not only the larvae that share a family resemblance—the tobacco hornworm moth is a dead ringer for the tomato hornworm moth. The names, though, give them away. *Manduca sexta*, the tobacco hornworm, has six pairs of orange spots (hence *sexta*) on its abdomen, while the

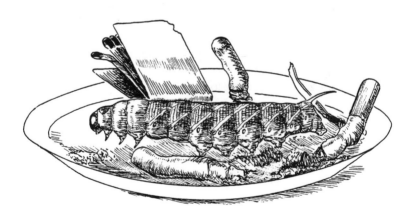

Tobacco hornworm

tomato hornworm, *Manduca quinquemaculata* (from the Latin for "five spots"), has (you guessed it) five pairs of orange spots on its abdomen. Otherwise, the moths are greyish to brownish in color and variously streaked with black stripes.

The family responsible for the family resemblance between *M. sexta* and *M. quinquemaculata* is the Sphingidae, the hawk moth, or hummingbird moth, family. Most members of the family share the same general gestalt—large and heavy-bodied with long pointed wings, long pointed abdomens, and a long tongue (or proboscis) that coils up in a spiral when not buried in the nectary of a flower. The hawk, or hummingbird, moths (as the names imply) are excellent fliers and, like hummingbirds, are capable of hovering in one place while feeding on a flower.

Adult moths emerge in the spring after overwintering as pupae three or four inches down in the dirt. At nightfall, female moths find a vertical surface to climb on and release an airborne chemical sex attractant, or pheromone, to bring in the boys. Mating takes place end to end on the vertical surface, with the female uppermost, and takes the better part of the evening. After mating only once, females then lay pale green spherical eggs one at a time on choice foliage. After three to eight days, the larvae hatch and begin chewing tobacco, tomatoes, or whatever other solanaceous plants stand in the way of their mandibles. After five or, occasionally, six molts, the larvae stop feeding, dig down into the soil, form a chamber, and pupate. Depending on latitude, there may be anywhere from one to four generations each year.

Tobacco (*Nicotiana tabacum*), as most people know, is full of nicotine, a neurotoxic chemical for which many addicted Americans will gladly pay over a dollar a pack. But nicotine has been used as an insecticide for about as long as it's been used for smoking pleasure. Exactly how the tobacco hornworm copes with a chemical that knocks out most of its colleagues has been a puzzle for years. One theory is that the gut of the tobacco hornworm keeps the nicotine in its ionized (or charged) form. In this form, the chemical can't penetrate the nervous system of the insect to do its damage. Thus, there is no compelling neurological reason for a tobacco hornworm to kick its habit (although, with six legs, it's well equipped to do so).

Chapter 4

Farm Friends

Alfalfa weevils

Blister beetles

Click beetles

Corn leaf aphids

European corn borers

Grasshoppers

Greenbugs

Green lacewings

Leafhoppers

ALFALFA WEEVILS

Every spring, there are a few industrious insects that aren't "making hay as the sun shines"—they're eating it instead. Chief among these is *Hypera postica*, the alfalfa weevil. Unlike its close relatives (such as the clover leaf weevil, clover root curculio, clover root borer, and the sweet clover weevil), the alfalfa weevil much prefers alfalfa to clover or other plants in the Leguminosae, the legume family. In fact, they came all the way over from Europe around the turn of the century to try alfalfa, American style.

Generally, the adults (greyish brown in color with wide blackish stripes down the back) pass the winter in hibernation in the crowns of alfalfa and in and around other leaf litter. Come spring, around April, they begin their alfalfa assault in earnest. Females use their elongated snout (typical of all weevils, but only medium-sized in *Hypera postica*) to hollow out a cavity in an alfalfa stem. They then turn around and insert anywhere from one to over four hundred lemon yellow oval eggs. Depending on the temperature, eggs hatch in four days to three weeks. Occasionally, eggs laid in the fall may not hatch for over five months.

The newly hatched larvae are yellowish in color but gradually assume a pale green color interrupted by a conspicuous white stripe down their back and a pair of faint stripes on the side. These plump, legless grubs feed for a short time inside the stalk but soon move out to mow down the opening leaf buds at the top of the plant. This relentless pruning has the effect of inducing lateral growth below the tip, which the weevil grubs focus their attention on after the leaf buds have been ravaged. There's very little of the alfalfa plant that isn't eaten—adult alfalfa weevils will even nibble on the woody stems if all the good stuff is already gone.

The larvae go through three to four molts, after which time a heavily infested alfalfa field looks almost bleached; both larvae and adults can skillfully excise the green parts of foliage, leaving inedible leaf skeletons

Alfalfa weevils: "See no weevil, hear no weevil, speak no weevil."

behind. Around mid-May, the mature larva drops to the ground and spins a lacelike spherical cocoon in which to pupate. Ten days later, the adult emerges and spends summer and fall eating alfalfa whenever the urge hits.

Overall, the alfalfa weevil is no friend of the alfalfa farmer. Fortunately, the farmer has a few allies in the trenches. For example, alfalfa weevils are frequently parasitized by a wasp, *Bathyplectes curculionis*, which lays eggs and completes development inside the weevil grub. In some areas up to 90 percent of the grubs die from parasitism. There is also a highly specific fungus which infects the weevil grubs in scattered parts of their range, turning them into pulpy mush. One of the most effective methods of controlling alfalfa weevils is judicious cutting of the alfalfa hay crop. If cutting is properly timed, hay can be harvested before too much damage is done, and, if the hay is removed from the field, the hot sun will take care of the sensitive eggs and larvae on the ground. Also, cultivating and dragging the fields after harvest isn't exactly a picnic for the weevils either. For advice on the proper timing of cutting, call your local county extension office; their agents will help you see no weevil, hear no weevil, and speak no weevil.

BLISTER BEETLES

When you were a kid, did you ever have trouble deciding what to be when you grew up? Blister beetles undergo a series of identity crises as a matter of course while growing up. Beetles in the family Meloidae begin life as an egg underground. The eggs hatch into long-legged frenetic larvae, called triungulins, who go about the business of finding food. In some species this consists of locating a grasshopper egg case in the soil and chewing into it. Other species feed on the eggs and larvae of bees, which are a little more difficult to find. These triungulins depart from the soil, find a flower, and climb up to the top of it. There they wait for a bee to land. When and if one does, they latch on and allow the bee to transport them to their new home.

Once established in their respective abodes, the triungulins molt into second-stage larvae with shorter legs and a stouter body. This caraboid larva resembles the immature stages of ground beetles (members of the family Carabidae). After several days of eating, the caraboid larva changes shape again and molts into a scarabeiform larva—a fat, C-

Blister beetle: "Look, buddy, you want hair or not?"

shaped grub with tiny legs resembling the larvae of scarab beetles (members of the family Scarabeidae). Two more molts, and the legs and mouthparts grow progressively smaller until, after a fifth molt underground, the larva is legless, mouthless, and covered with a thick dark exoskeleton.

This coarctate larva, or pseudopupa, is in the developmental stage that overwinters (which is just as well, since it has no mouthparts to eat with). Come spring, the pseudopupa molts again and turns into an active larva which has mouthparts but doesn't feed and has legs but can't use them. This scolytid larva, resembling in appearance the off-spring of bark beetles (members of the family Scolytidae), prepares a chamber in which to pupate and molts again, this time into a true pupa. After about two weeks the adult blister beetle emerges and is content to remain a blister beetle for the rest of its life.

Blister beetles are pretty distinctive, as beetles go. They're generally soft-bodied, short-winged (their abdomen is visible from above), and "long-necked" (although no insect has a real neck, blister beetles appear to have one, since the first three body segments behind the head are constricted). Many blister beetles are brightly colored, with iridescent black, purple, green, or brown as popular options.

Putting larval habits behind them, adult blister beetles eat plants instead of insects and occasionally make a nuisance of themselves in the process. The striped blister beetle, *Epicauta vittata*, is sometimes called the old-fashioned potato beetle, because it's been damaging the foliage of the potato since before the Colorado potato beetle got into the business. The black blister beetle, *Epicauta pennsylvanica*, which can sometimes be spotted basking and eating pollen on goldenrod flower heads, is partial to many ornamentals, including chrysanthemum, asters, gladiolus, and zinnias. Both species feed on alfalfa (blossoms and leaves) and can spoil hay for cattle and horses if they're baled in along with the plants.

Blister beetles (also sometimes called oil beetles) are called blister beetles because they release an oily substance which contains cantharidin. Cantharidin is a vesicant, a substance that raises a large, painful blister when it comes into contact with human skin. For centuries, cantharidin, as contained primarily in the bodies of the meloid *Lytta vesicatoria*, or Spanishfly, has been used for a tremendous variety of medicinal and not so medicinal purposes. Spanishfly (or cantharis, a preparation of the dried beetles) has been prescribed over the centuries for almost every imaginable illness, including rabies, rheumatism, gout, carbuncles, snake bite, lice, earache, and painful urination. According to the classification system of Galen, Spanishfly is "extremely hot,"

and it was frequently used as an aphrodisiac. Unfortunately, cantharidin is not only extremely hot, it's extremely toxic. As little as thirty
milligrams can kill a person, which could certainly put a damper on a
romantic evening. Today cantharidin is licensed only for veterinary
medicine and is used primarily as a counterirritant and vesicant. At
one part per thousand in solution, however, it's been found to stimulate
hair growth—so blister beetles, while they may not really be an aphrodisiac, are at least hair-raising.

CLICK BEETLES

Insects have devised a variety of modes of locomotion—they fly, crawl, slither, squirm, wriggle, and hop, to name a few. The click beetles, though, in the family Elateridae, have developed a rather distinctive means of moving about: they flip, or rather, click. Unlike most insects, click beetles have a jointed flexible thorax (the body region to which the legs are attached). A long spine projects from one section of the thorax and fits into a groove in the adjacent section. If a click beetle happens to land on its back, it arches its back, juts its thorax into the air, and then slaps it back onto the ground. The spine clicks into the groove and the force propels the insect into the air, flipping end over end. Like as not, the beetle lands on its back again, and it repeats the process until it eventually ends right side up and can run away. All this goes on in spite of the fact that click beetles have two pairs of eminently serviceable airworthy wings and three pairs of perfectly workable legs (you figure it out).

Adult click beetles are easily recognizable. They are distinctively elongate with parallel sides and slightly rounded head and tail ends. They range in color from speckled grey to dingy brown-and-black. One species is not to be missed—the eyed click beetle, *Alaus oculatus*. It

Click beetle and wireworms: "Excommunicated? I didn't even know I was Catholic!"

reaches an inch and a half in length, and its thorax is adorned with two very large eyespots which stare balefully. One Central American species even has eyespots that glow green in the dark. Most click beetles, though, are a more conservative color and length, averaging less than an inch from head to tail. Adults can be found on flowers, vegetation, and tree bark.

It's unclear what click beetles eat as adults, or if indeed they eat at all, for a starved female can lay as many eggs as a well-fed one. From late May through July females lay eggs in the soil, from two to four hundred all told, and the immature click beetles, called wireworms, emerge. Wireworms in no way resemble their parents—they're long and reddish white, with a tough, shiny exoskeleton. They are also voracious eaters. While a few exceptional species are predaceous, most feed on roots, fibers, seeds, humus, and anything else they can find in the soil.

Mature wireworm larvae attack roots and tubers of a number of garden crops, including beets, beans, corn, potatoes, and various weeds. Immature larvae are particularly partial to seeds, and around early spring they might well be the reason that garden seeds seem to disappear as soon as they're planted. Wireworms cause a great deal of economic damage each year to corn and wheat fields, where they hollow out the seeds as they are planted and chew through leaf sheaths into the stems of the seedlings that do manage to germinate. A serious infestation can reduce the yield of wheat by as much as twelve bushels per acre.

Controlling wireworms is, to say the least, challenging. Again, unlike most insects which reach maturity and reproduce with breathtaking speed, wireworms are slow to mature. It takes, on average, three to eight *years* and ten to twenty-four molts for a wireworm to grow up; if you had wireworms last year, most likely the same ones will be back this year. Many different approaches have been taken to eradicate these insects. Cultivating in late July will crush the pupae and expose the larvae to the elements, and clearing unharvested plants will starve them out. Another approach is to ensure the rapid and early germination of seeds, so they can escape wireworm attack by outgrowing their attackers.

Perhaps the most original approach to wireworm control was instituted by the Bishop of Lausanne, Switzerland, in 1479. According to F. Cowan in his 1865 book, *Curious Facts in the History of Insects,* upon hearing of the ravages wrought by wireworms in the fields, the Bishop had the brilliant idea of excommunicating the wireworms. The worms were duly brought to court (in absentia), found guilty, and expelled from the church. When the wireworms returned to continue their depredations the next year, the bishop declared a mistrial. The verdict had been invalid since no friendly witnesses—like the birds that eat wireworms—had been called to testify in their defense. This method of control for wireworms, not surprisingly, hasn't been widely used since.

CORN LEAF APHIDS

Leave it to the insects to carry anything to its logical extreme. The corn leaf aphid has added new dimensions to the concept of feminism.

The corn leaf aphid, *Rhopalosiphum maidis*, is a major pest of corn, barley, and sorghum wherever it occurs: that is, everywhere in the world between the latitudes 40° north and 40° south. *Rhopalosiphum* is tiny (about one-twentieth of an inch long at maturity) and fairly nondescript as aphids go. It is greenish blue with a dark green head, with or without wings, and most frequently is found in the heads and young tassels of corn. Of course, like the rest of the aphids, the corn leaf aphid has long piercing mouthparts which it uses to drain vital fluids from its host. Infestation of corn by the corn leaf aphid is associated with malformed kernels, reduction in the number of kernels, and even barrenness (failure to produce an ear at all). To add insult to injury, the corn leaf aphid can also spread the viral disease maize dwarf mosaic and reduce yields still further.

Corn leaf aphids

Rhopalosiphum maidis is found in every state in the union, but it sensibly spends its winters in the sun belt. Come spring, aphids arrive en masse to begin the business of breeding in earnest. One factor that contributes to the corn leaf aphid's prowess as a pest is its tremendous powers of increase. In fact, this *R. maidis* can do without the benefit of mating or, for that matter, males. No wonder, then, that male corn leaf aphids are extremely hard to come by. In one intensive ten-year study in Arizona from 1914 to 1924, two investigators managed to find only forty males. Five more males were found in Kenya in 1953, and since that find, they've been few and far between.

The females' production of offspring without mating is known by the cognoscenti as parthenogenesis. The particular sort of parthenogenesis practiced by *R. maidis* is called thelytoky, which is the production of female offspring by female parents. In addition, *R. maidis* adds another twist to its reproductive life: it (or she) dispenses with the egg stage altogether and gives birth to fully formed baby aphids. While live birth is no great shakes for us warm-blooded mammals, it's still an unusual feat among insects.

It takes only five to six days for a newborn aphid to reach adulthood (if it survives beating rains, fungus diseases, cold temperatures, and a host of insect predators and parasites), and after another one to three days it's ready to produce up to one hundred more aphid young. All of these factors—short development time, parthenogenetic reproduction, and production of many live young—add up to a tremendous capacity for increase. In some parts of the country the corn leaf aphid can produce up to fifty generations per year (and remember that's one hundred aphids each generation), making them among the most prolific of all insects. The corn leaf aphid might well have served as the inspiration for the thought, "There's a sucker born every minute."

EUROPEAN CORN BORERS

While most living things try to keep warm all winter, *Ostrinia nubilalis*, the European corn borer, never loses its cool. A voracious caterpillar in the family Pyralidae, it spends its summers vigorously devouring the leaves, stems, and stalks of over two hundred species of plants, including asters, beans, corn, dahlias, and straight on through the alphabet to zinnias. Around September, however, the European corn borer slows its pace—it stops feeding, halts its development in the fifth larval stage (or instar), and drops its oxygen consumption to about 25 percent of what it had been during the summer months.

This state of suspended animation is called diapause. As temperatures fall with the onset of winter, the European corn borer's temperature falls right in step with the season. The corn borer can lower its body temperature without freezing at this time of year much more effectively since it accumulates glycerol, a chemical similar in both appearance and function to ethylene glycol, the antifreeze in car radiators. Glycerol can only go so far, however; beyond that point, European corn borers are perfectly capable of freezing solid and still surviving.

In the spring, these caterpillars thaw out and spin a casual cocoon inside their tunnels. There they pupate, and after several weeks they emerge as adults. The corn borer moth, with a wingspan of an inch at best, is brownish beige in color with dark irregular bands running across its wings. The males are conspicuously darker than their mates. When males and females want to get together, females release a blend of chemicals at night from their abdominal glands which acts as a sexual

European corn borers: "Poor guy . . . looks as if he forgot to change his anti-freeze."

attractant or pheromone for the males. Once the business of mating is dispensed with, the females begin to lay up to five or six hundred eggs in clusters of about one to four dozen, particularly on the undersides of leaves.

Although the European corn borer will usually eat just about whatever is put in front of it, most of its bad press comes from the fact that it is particularly partial to corn. On a corn plant, after the eggs hatch in a week or so, the young larvae (cream-colored to pink, with a brown head and two brown spots on the top of each abdominal segment) feed on the leaves, in the tassels, beneath the husks, or in the stem between the stalk and the ear. After two or three molts, they move into the stalk and, start boring. Their activity is anything but boring to farmers, who sometimes watch the borer population increase to the point that corn plants collapse from the internal demolition job going on; some fields average over twenty borers per stalk. The estimated loss of grain corn to European corn borers in 1969 was 163 million bushels, worth $183 million at the time.

The European corn borer has made its way to the top of the pest list in the U.S. rather quickly, since it was unknown in this country until 1917. Experts believe the first immigrants arrived in shipments of broom corn from Hungary or Italy around 1908 or 1909. It's clear, however, that other corn borers snuck in at other times; the American version of the European corn borer is actually a conglomerate. Soon after it was discovered in the U.S., discrepancies in its biology began to appear. In New York, borers only produced one generation per year; in Massachussetts, they managed two. Even when Massachusetts borers were raised in New York, they still underwent two generations per year. In New York and New Hampshire, by 1925, both types of borers were found, but that same year, the borers found in Ohio and Michigan were one-generation types. Illinois borers, first reported in 1939, were a mixed bag: about half of the first generation stayed in diapause, and the other half went on to start a second generation.

The working hypothesis to explain all this variation is that there are at least two genetic strains of European corn borer more or less randomly scattered over the corn-growing states. These strains differ not only in their seasonal biology but in their sex life as well. Females of different strains produce different blends of pheromones which attract only males of their own genetic background. These different strains probably are descendants of immigrants from different places in Europe, which went their separate ways upon arrival in the U.S. Like it or not, the American "melting pot" also seems to contain a few insects.

GRASSHOPPERS

The name "grasshopper" is, for the most part, an exercise in wishful thinking. If only grasshoppers had restricted themselves to grass, considerable human suffering through the ages might well have been averted. As it is, however, grasshoppers cause more direct injury to crop plants than any other kind of insects, and through the centuries they have routinely left famine and economic ruin in their wake. Part of the problem is that most grasshoppers eat a tremendous variety of plants, everything from grasses and other small grains to vegetable crops, from range plants to fruit trees. The other part of the problem is that most grasshoppers eat a lot of plants, period. A "light" infestation averaging 6 or 7 grasshoppers per square yard can eat as much alfalfa hay as a cow; at 15 or 20 per square yard, they can knock off a ton of hay in a day. And a heavy infestation can average four to eight bushels of grasshoppers per acre, with 200,000 grasshopper per bushel (and that, as they say, ain't hay).

Grasshoppers: ". . . and after we ravaged Iowa, we decided to visit you folks here in Illinois . . . say, are you going to finish that sandwich?"

There are over six hundred species of grasshoppers in North America, but only about five species cause 90 percent of the grasshopper damage to cultivated crops. The largest of the crew is the Carolina grasshopper, *Dissosteira carolina*. While not as destructive as some, it's one of the more visible grasshoppers, not only by virtue of its size (over two inches in length at maturity), but by virtue of the fact that it frequents roadsides, railroad tracks, and other heavily trafficked areas. When disturbed, it readily takes to wing (that wing being a rather conspicuous black with a yellow border).

The differential grasshopper, *Melanoplus differentialis*, at maturity hops out at 1½ inches in length. Besides its size, its most distinguishing feature is the black chain of chevrons on its hind legs. The two-striped grasshopper, *Melanoplus bivittatus*, is 1 to 1½ inches in length when full-grown and can be distinguished by the yellow stripe extending from head to wingtip on each side of its body. Like the differential grasshopper, the two-striped grasshopper is partial to lush vegetation, and in dry seasons is restricted to drainage ditches and streambeds. The American grasshopper, *Schistocerca americana*, has brown spots and a yellow belly. It's often called a "bird grasshopper," due to its impressive aerial skills.

The two most destructive species, the red-legged and the migratory grasshoppers, barely reach an inch in length at maturity. The red-legged grasshopper, *Melanoplus femurrubrum*, is brownish red with a distinct pinkish cast to its hind legs. It is oftentimes a serious pest in soybean fields, cutting through the pods and stripping vegetation. The migratory grasshopper, *Melanoplus bilituratus*, is perhaps the most serious and most widely distributed grasshopper species. It thrives equally well on sparse native vegetation and on luxuriant waves of grain. In appearance it resembles the red-legged grasshopper, with less of a pinkish color to its legs.

The migratory grasshopper, like most grasshoppers, lays a cluster of eggs in the ground encased in a cemented mass called a pod. In the case of the migratory grasshoppers, the pod contains about sixteen to twenty eggs. The eggs are laid September through October and hatch the following May through July, usually after spring rains. The nymphs (or immature grasshoppers) look, for the most part, like the adults, only smaller and without wings. After four to six molts the nymphs reach adulthood (in four to seven weeks, depending on the weather).

All grasshoppers move a great deal while feeding, from field borders into fields and from one field to another, but if pickings are slim when they acquire their wings, migratory grasshoppers take off en masse and really move, up to six hundred miles at times. The last great migration of this grasshopper took place from 1938 to 1940 in the Northern Great Plains states.

Still, that plague didn't compare in magnitude to the outbreak of the Rocky Mountain locust, *Melanoplus spretus*, in the 1870s. Properly speaking, a locust is simply a migratory grasshopper and, while no close relative of the plague locust of biblical fame, *M. spretus* did OK on its own. From the high plains of Colorado and Montana east to Nebraska, the Rocky Mountain locusts consumed virtually every shred of green vegetation. They literally darkened the sky for hours as they passed overhead. A bounty was placed on their tiny heads (one dollar a bushel for hatchlings or five dollars a bushel for egg pods), and in Kansas in 1877 the Grasshopper Act was passed, stipulating that every able-bodied male between the ages of twelve and sixty-five had to be prepared at all times to collect locusts when ordered to do so.

By 1878, however, their numbers began declining, and by the turn of the century *M. spretus* virtually dropped out of sight. The last living specimen was sighted in Manitoba in 1902. Their disappearance remains an entomological mystery—the swarms of Rocky Mountain locusts passed into oblivion like the great western herds of buffalo, but without even the flipside of a nickel to commemorate their passing.

GREENBUGS

If you're ever wondered why the grass isn't always greener on the other side of the fence, it may be that there's an infestation of greenbugs there. The greenbug, *Schizaphis graminum*, is a tiny pale green aphid with an appetite for any plant in the grass family, including wheat, oats, corn, rye, rice, barley, grain sorghum, spelt, and many wild prairie forage grasses as well as cultivated grasses (according to one source, S. J. Hunter in 1909: "You see this shifty species is not likely to want for food").

Greenbugs aren't content merely to suck the sap of the plants on which they feed. As they suck, they also inject a poisonous saliva that breaks down healthy plant tissues and causes discoloration and death. Hence, the most characteristic symptom of greenbug damage (despite its colorful name) is the presence of small red or yellow spots on grass blades. As the greenbug population grows, as it is wont to do, the discoloration grows with it, until fields of previously green grain are interrupted by large circular patches of brown dead plants surrounded by a ring of bright yellow plants. Eventually, these areas can coalesce, and the proverbial amber waves turn brown and die.

Schizaphis graminum inflicts damage out of all proportion to its size. At maturity, a single aphid rarely exceeds one-sixteenth of an inch in length, yet this species regularly causes a 1 to 3 percent loss of the annual world wheat crop and has at times reduced the southwestern U.S. wheat crop by as much as 25 percent. What greenbugs lack in size they more than make up for in numbers.

Greenbugs

In the northern U.S., their black shiny kidney-shaped eggs hatch in early spring to produce a generation of pale green females. These females, without benefit of mates, begin to produce young (like other aphids, greenbugs skip the egg stage and produce living young). Within one to two weeks, these offspring, all female, can then produce their own young. When cold weather hits, five to fourteen generations later, males appear in the population. These males mate with females, who then lay eggs on the leaves of dying grasses for overwintering. In the South, where cold weather never hits, eggs are dispensed with altogether, and greenbugs reproduce for thirty or more generations.

Fortunately for the blades of grass of the world, the life of the greenbug has a few rough spots. Immature ladybugs can eat up to one hundred greenbugs a day; then there are lacewings and syrphid fly larvae which also find them easy pickings. *Aphidius testaceipes*, a wasp parasite, lays its eggs inside the body of the aphid, where the hatching grubs eventually eat the aphid from the inside out.

Recognizing a greenbug is no mean feat inasmuch as its appearance changes with the season. The aphids hatching in the spring grow up to be pale green and wingless. Summer greenbugs are either winged or wingless, pale yellow to bluish green, with a dark green racing stripe down their backs. The wingless, egg-laying, fall-form females are more or less similar to the summer form. One characteristic feature is that the cornicles (paired structures projecting from the rear of the abdomen like tail pipes) are green with black tips. The winged males resemble the winged females but are only about one-twentieth of an inch in length. To complicate matters, there are several other aphids frequenting plants in the grass family with more than a passing resemblance to greenbugs. These include corn leaf aphids and yellow sugar cane aphids. One thing is certain, however; if it's feeding on a plant outside the grass family, it's probably *not* a greenbug. Despite any other flexibilities in its life history, the greenbug is fanatically devoted to corn, rice, wheat, sorghum, and other grasses. To feed on anything else would go "against the grain."

GREEN LACEWINGS

Some of the greatest farm aids around are species in the order Neuroptera, specifically, the green, brown, and dusty lacewings. While members of the order all share the characteristic two pairs of gossamerlike wings crisscrossed with numerous veins and crossveins (whence cometh "Neuroptera" or "nerve wing"), they differ wildly in their habits. In fact, some of the Neuroptera can be classified as true insect eccentrics, like the spongillaflies, which are parasitic on freshwater sponges, or the mantispids, which live inside the egg sacs of spiders. The lacewings, though, are insect predators extraordinaire.

The green lacewings, in the family Chrysopidae, are particularly partial to aphids, mites, and other small agricultural nemeses. Their appearance is deceiving, however. The adults are fragile, delicate, effete-looking creatures with a relatively small head. They're generally pale pastel colors, including lime green or lemon yellow, and one common species, *Chrysopa oculata*, even has large eyes of burnished gold. To top it all off, they're weak, awkward flyers, and some produce an offensive odor. Altogether, they don't present an image typical of a vicious, ravenous predator. From the aphid point of view, however, they are lean green killing machines.

Green lacewing (in larval stage, as aphidlion)

It's the larvae (or immature stages) of the Chrysopidae which stalk and kill other insects. The lizardlike, spindle-shaped larvae, mottled greyish in color and equipped with tufted tubercles and elongated sickle-shaped mandibles, are so voracious they're known as aphidlions. The larvae undergo three molts in five to six weeks, depending on the supply of aphids, and attach to a tree trunk (or other convenient location) a spherical silken cocoon in which they pass the winter in hibernation. Come spring, they cut a hinged lid at the top of the cocoon and begin a new round of aphid slaughter by mating and laying eggs. Lacewing eggs are difficult to mistake for any other kind of insect egg; although otherwise undistinguished, the oval white egg is deposited on the end of a long thin filament attached to a leaf. Presumably, the filament keeps the egg out of the reach of roving gangs of egg predators searching for sustenance on leaf surfaces.

The lives of the other lacewings are more or less similar. Brown lacewings, in the family Hemerobiidae, are more often (as the name implies) brown than green, and the larvae are known as aphidwolves instead of aphidlions. Both larval and adult stages eat aphids, scales, and other small arthropodan food items. And the dusty wings, in the family Coniopterygidae, differentiate themselves with a whitish mealy (or farinaceous) powder on their wings.

Wrestling tiny soft-bodied aphids or mealybugs to the ground may not seem like such an impressive feat of predation, but the major obstacle the lacewings encounter is not resistance from their prey per se but from the pugnacious insect bodyguards associated with their prey. Aphids, scales, and mealybugs produce honeydew, a favorite food of ants, and many ants zealously protect their providers from marauding predators. Lacewing larvae have devised some remarkable ways of getting by the guards. Some attach bits of dust or lichen to their backs to look as much like nothing in particular as possible. One species which feeds on woolly alder aphids sneaks up and plucks off the waxy filaments from the aphid's back and attaches them to special grappling hooks on its own back. Thus disguised, the now-woolly aphidlion can move in and among the ant bodyguards without detection. The fable, then, falls a little short of the mark—this is clearly a case of a lion, and not a wolf, in sheep's clothing.

LEAFHOPPERS

Labor Day weekend is, for most people, a last fling at summer before weather-stripping, insulating, and other winter preparations begin in earnest. Most leafhoppers, however, rarely ever concern themselves with such matters. Instead of toughing it out by hibernating through the winter months, leafhoppers, as a rule, spend their off-season down south in warmer climates. Come fall, some species travel in large swarms that fly over two hundred miles to reach a suitable spot to wait out the cold weather. Then, in the early spring, adults fly north to mate, lay eggs, and avail themselves of what America's heartland has to offer.

One of the more familiar leafhoppers, the potato leafhopper, *Empoasca fabae*, is thought to migrate only once, from south to north each spring. It can feed on over one hundred plants, including such crop plants as eggplant, alfalfa, beans, soybeans, celery, clover, apples, and (of course) potatoes. Like all cicadellids, the potato leafhopper feeds

Leafhoppers

by tapping large veins on the undersides of leaves and sucking plant juices through its tubelike mouthparts. Leafhoppers cause considerable economic damage not only because they reduce the vigor of plants by sucking out vital fluids, but also because in the process they inject salivary substances that block the vascular system of the plants and cause wilting, curling, discoloration, and stunting, often called "hopper burn." In addition, they act as vectors for many plant diseases, not the least of which are aster yellows, curly top, Pierce's disease, purple top wilt, tomato yellows, and potato yellow dwarf.

Empoasca fabae is fairly representative as leafhoppers go. About one-eighth of an inch in length, it's mid-sized; pale green with white spots on head and thorax, it's not as flashy as some and a bit more striking than others. All leafhoppers have a characteristic, easily recognized wedge-shaped appearance and a not so easily recognized double row of spines running the length of the hind legs.

Empoasca fabae arrives in its northern breeding grounds in early spring. Females lay eggs along the main veins of leaves, so the emerging nymphs won't have far to travel in search of a meal. After five molts and about two weeks, the process starts in all over again and repeats itself anywhere from two to five times, depending on latitude, before the adults in the fall either die off or head south.

Leafhoppers escape from their potential enemies (as their name implies) by hopping, often for great distances. They also have the peculiar habit of running away sideways, enough to throw all the more conventional predators off balance at first. Even more peculiar is the habit to which sharpshooters, a group of rather large-sized leafhoppers, owe their name. Generally more brightly colored than the other leafhoppers, favoring stripes rather than spots and reds and blues instead of greens and browns, sharpshooters feed in much the same manner, by piercing the veins of leaves and sucking the sap. But after they've extracted all the nutrition they need from the fluid, they expel the remainder through the tip of their abdomen with great force and at some distance. One redbanded leafhopper was observed to expel one droplet a second for two straight minutes without missing a shot, a pretty tough record even for human sharpshooters to beat.

Chapter 5

Roadside Attractions

Crab spiders

Goldenrod ball gall flies

Golden tortoise beetles

Monarch and viceroy butterflies

Painted ladies

Parsnip webworms

Praying mantis

Sulfur butterflies

Woollybears

CRAB SPIDERS

When Sir Walter Scott penned "Oh what a tangled web we weave, When first we practice to deceive," he most certainly didn't have crab spiders in mind—crab spiders regularly practice deception without spinning a single strand of silk.

Crab spiders, species in the family Thomisidae, owe their name to an imagined resemblance to the marine-variety crab, mostly due to the fact that the first two of their eight legs are enlarged and pincerlike compared to the rest. Even though, like real crabs, crab spiders are perfectly capable of moving sideways or even backwards with considerable facility, they are not likely to deceive anyone into thinking they are aquatic crustaceans. Deception only enters into the picture at dinnertime; crab spiders are ambush predators who conceal themselves and lie in wait for unsuspecting morsels to walk by unawares. When a tidbit strolls within range, the viselike forelegs snap shut to immobilize the prey while the crab spider alternately pumps digestive enzymes into the body and sucks out its liquified contents.

Crab spiders: "There's nothing sadder than a color-blind crab spider."

The technique of hunting by ambush is widespread throughout the animal world (and is even resorted to by duck hunters and other human beings). But crab spiders carry it one step further. As spiders go, thomisids are almost flamboyant; *Misumena vatia*, the common flower crab spider, runs the spectrum from creamy white to saffron yellow and is usually adorned with a row of red spots down the middle of its body or with a series of black lines along its sides. The outfit, however, is part of the strategy. The flower spider sets up shop on flowers with petals that match or complement the pattern on its body. The most common flowers in the fields and roadsides inhabited by the spiders are flowers in the daisy family and are almost exclusively yellow or white. With alarming accuracy, yellow spiders sit and wait on yellow flowers, white ones on white flowers.

Crab spiders can't be given too much credit for cleverness. They contain a pigment that is sensitive to white or yellow reflected light and that changes according to background color. So if a white spider sits on a yellow flower, within one to twenty days it will be a yellow spider on a yellow flower. The color change is reversible, should the need arise for a spider to show its true colors. Once a spider matches its background, it's very difficult to see. The tiny spider, only one-third to one-half inch in length, can thus use the element of surprise to ambush prey the size of bumble bees and butterflies. Moreover, crab spiders are less likely to be spotted by their own enemies.

Even with their camouflage, crab spiders aren't ones to flaunt their invulnerability (or their invisibility). For the most part they are reclusive and spend much of the day pressing their flattened bodies into plant, rock, or soil surfaces or hiding in cracks or under debris. They move about a great deal, the only exception being at egg-laying time, when females guard their silken lens-shaped egg sacs concealed on leaf surfaces until the spiderlings hatch out.

There are over one hundred species of crab spiders in the United States alone. While not all of them can change their colors in order to disappear into the woodwork, most depend on looking invisible for both hunting purposes and protection from enemies. Some resemble bird droppings; others resemble fruits, dried seeds, leaf buds, or even roses. So, as poets go, Gertrude Stein as well as Walter Scott could learn a thing or two from crab spiders.

GOLDENROD BALL GALL FLIES

It takes a lot of gall to get through life if you're an insect, but it takes more for some than for others. Take *Eurosta solidaginis*, for example. *Eurosta solidaginis* is a member of the family Tephritidae, the "true fruit flies," or picture-winged flies. Adults are chunky medium-sized flies, running up to about one-third of an inch in length, with a brownish red body and brown wings adorned with a network of fine lines and clear windowlike markings. For the most part, they're totally inoffensive to most life forms on the planet.

However, the female *E. solidaginis* every spring seeks out the stem of the goldenrod (a plant in the genus *Solidago*) and inserts an egg directly into the terminal bud. After hatching, the immature *E. solidaginis* is a legless pale white maggot which is entirely featureless but for a pair of dark strong mouthhooks at what passes for the head end. The young maggot uses the mouthhooks to tunnel its way down below the growing point of the plant, where it settles into a small chamber. The plant responds to this indignity by developing a special swelling around the maggot's chamber. This abnormal tissue growth is called a gall, and *E. solidaginis* is otherwise known as the goldenrod ball gall

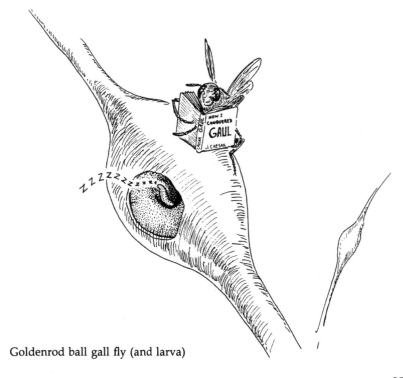

Goldenrod ball gall fly (and larva)

fly. It is believed that the gall insect essentially commandeers the plant's hormonal system in order to produce the characteristic swelling along with depositing a thick tissue layer (up to one-third of an inch, wall to wall) on which to snack. Swelling and gall formation are complete in three to four weeks.

The larva develops safe and secure in its gall until October, when the temperatures fall. It then enters diapause, a state of suspended animation, till spring. The goldenrod stems turn brown and dry but remain for the most part upright, so goldenrod ball galls are visible all winter long, each equipped with a refrigerated, motionless maggot inside.

When spring arrives, the maggot chews a tunnel out to the very edge of its gall and then returns to the central chamber to pupate. *Eurosta solidaginis*, like many of the so-called higher flies, produces a puparium—a thick, tough brown capsule that form-fits the maggot's body. In a couple of weeks metamorphosis is complete, and the adult fly first crawls out of the puparium through an escape hatch and then crawls out the prefabricated tunnel to the outside world. As far as goldenrods are concerned, being galled is galling; seed production drops up to 40 percent on stems with galls.

It all seems simple and straightforward, but for centuries galls puzzled some of the greatest scientific minds in history. Theophrastus in 300 B.C. thought the insect's presence in the gall was a coincidence. Francesco Redi, of all people (he's the man in the seventeenth century who disproved the idea that maggots appear spontaneously in meat), thought the insects appeared spontaneously inside. Still others thought plants extremely generous, if not terribly bright, for providing a home for a parasite.

Actually, the truth lies in between. The size and location of the goldenrod ball gall depends at least partly on the ability of an individual insect to induce the gall and partly on the ability of the individual plant to produce the gall. Another factor involved is the presence of animals that eat goldenrod ball gall residents. For example, small galls are easily penetrated by the ovipositor of a parasitic wasp whose grubs eat the gall occupant and comfortably ride out the winter in its place. Large galls, on the other hand, are easily spotted in winter by hungry chickadees and downy woodpeckers. So gall size is the result of the interplay between plant, insect, and predators. In other words, as Julius Caesar (otherwise not renowned for his entomological powers of observation) once remarked, "All gall is divided into three parts."

GOLDEN TORTOISE BEETLES

The standard proverb, "all that glitters isn't gold," is no less the case for *Metriona bicolor*, the golden tortoise beetle. As beetles go, *Metriona bicolor* is, if nothing else, striking. It owes both its common and scientific names to its unique color scheme. "Golden" refers to the brilliant shiny gold color of the adult beetle. *Bicolor*, on the other hand, refers to the beetle's ability, upon disturbance, to change its color from brassy or burnished gold to more subtle violet to subdued (almost bordering on drab) reddish brown in under one minute.

The tortoise beetle's quick-change act is nothing but simple physics. The beetle can, by altering the water content of its cuticle, change the ways in which light waves reflect and intersect, thereby changing its apparent color. The change is not only rapid, it's totally reversible, and the beetle can resume its classier colors when danger has passed.

Metriona bicolor, though, is not the only beetle that can glitter at will. Many members of the subfamily Cassidinae (or tortoise beetles) can change colors from variations on shiny bright or brilliant to dull and drab. The colors on the shiny end of the spectrum are so striking

Golden tortoise beetle: "I don't care what all the other tortoise beetle mothers let their kids do. Take it off before you come to the supper table."

that they may have been the inspiration for the name of the entire family of beetles of which the "gold bugs" are only a small part, the Chrysomelidae (from the Greek for "golden"), a family with over two hundred thousand species (the majority of which come in one standard color with no metallic options).

Aside from its golden color, the tortoise beetle is otherwise less than prepossessing. Under one-fourth of an inch in length, the roundish domed body of the beetle is vaguely reminiscent of that of a turtle—hence the appellation tortoise beetle. Adults spend the winter hiding under leaves and other litter and emerge in late spring to mate and lay eggs. Fertile *Metriona bicolor* lay eggs primarily on plants in the morning glory family, or Convolvulaceae. The same tastes in a related species, *Chelymorpha cassidea*, the argus tortoise beetle, make it an occasional pest on sweet potatoes, one of the more economically important members of the Convolvulaceae. While female golden tortoise beetles lay one or two eggs at a time and conceal them underneath a black pitchlike material, the argus tortoise beetle lays clusters of fifteen to thirty, each attached to the leaf by a stalklike projection.

If the adult beetles resemble tortoises, the larvae resemble nothing else ever seen before, at least on this planet. They are oval in shape with innumerable branched spines projecting around the perimeter of their flattened brownish bodies. At the nether end is a greatly elongated projection resembling a two-tined fork. The young larvae feed by scraping the leaf tissue, leaving behind thin windows of plant cuticle. Larger larvae, as well as adults, make neat circular holes in the leaves as they feed. The larvae proceed to grow and molt, and their shed skins accumulate along the anal fork, as does all their fecal material, or frass. By the time the larvae are ready to pupate, the anal fork is laden with a mass of excrement, silk, and shed skins that often outweighs the grub itself.

Though their habit of carting all their worldly (but less than precious) possessions around with them has earned them the nickname "trash peddlers," tortoise beetle grubs aren't selling anything. A grub uses this revolting mass to hide; when it's motionless it looks thoroughly unappetizing to most potential predators. If a pushy predator persists in disturbing it, it then waves its fork in the appropriate direction and strikes its enemy with its mass of frass, at which point ants and other small predators have been seen frantically running away to groom and clean themselves. It seems incongruous somehow that the beetle that most resembles a precious metal begins life as a grub that resembles a city landfill, but there's a lesson to be learned that entomologists know well—it's not easy to separate the gold from the dross, or the gold bug from the frass.

MONARCH AND VICEROY BUTTERFLIES

The vast majority of insects generally go out of their way to keep a low profile and avoid notice; the monarch butterfly, however, is literally a conspicuous exception. The adult is the highly visible, very familiar orange butterfly with black-bordered, white-spotted wings almost four inches across. The bright green caterpillar of the monarch butterfly is rather ostentatiously striped with black and yellow bands.

There's a good reason, though, for what would otherwise appear to be ill-advised brazenness; monarchs embody the principle "you are what you eat." As a caterpillar, the monarch (*Danaus plexippus*) feeds exclusively on milkweeds—plants in the family Asclepiadaceae. These are brightly colored wildflowers that grow along roadsides and in abandoned lots and fields throughout much of North America and Canada. The seemingly innocuous milkweeds are laden with toxic chemicals called, variously, cardenolides, cardiac glycosides, or heart poisons. The "heart" part of the name (*cardia* in Greek) derives from the fact that these chemicals are potent heart stimulants and can induce cardiac irregularities if consumed. They have yet another insalubrious (not to mention potentially socially embarassing) property as well—they cause emesis (a medically polite term for "losing one's lunch"). These chemicals are enough to deter most insects from feeding on the leaves of

Monarch butterfly

99

milkweeds, but the monarch and its close relatives are immune to the effects of the toxins. What's more, they go one better and, as caterpillars, store the cardenolides in their bodies for their own protection. When an insectivorous bird grabs a monarch caterpillar, it immediately wishes it hadn't. It gets a mouthful of heart poison and, if it tries to down the caterpillar, it just throws up. When the caterpillar pupates (after five molts, and two to three weeks after hatching from its egg), it retains the stored heart poisons, and as a result the adult butterfly is also protected. In one study, one poor blue jay threw up nine times in only thirty minutes after consuming just a single butterfly. Now, birds (even from a human perspective) are not that bird-brained, and most eventually learn to avoid eating monarchs under any circumstances—hence, the leisurely, casual flight of the adult and the bold behavior of the caterpillar. They are secure in their nauseating properties.

Monarchs are among the earliest butterflies to appear in the spring, and, just like some of their human counterparts, they go south for the winter. Great masses of them depart eastern and central North America as the weather cools and gather in flocks on their way south. The monarch is one of the strongest flyers among butterflies and has been found flying over five hundred miles from shore. The butterflies head for several clearly demarcated overwintering spots, including such popular human vacation spots as Mexico and the Monterey Peninsula in California. Literally millions gather at these spots. Each spring, the survivors make a return flight back to their northern summer homes to lay eggs. Their offspring spend the summer munching on milkweed, nauseating birds, and entertaining butterfly watchers.

Butterfly watchers must be advised, however, to accept no substitutes. Not every orange-and-black butterfly can give you heart trouble. The viceroy butterfly, for example (going by the name *Limenitis archippus* in entomological circles), is a dead ringer for the monarch—same reddish-orange color, black borders, and white flecks. Even seasoned entomologists confuse the two species, as do most of the major insect predators in the neighborhood. Any vertebrate hapless or hungry enough to sample a monarch is rewarded with retching for its efforts, and after such an experience it usually learns to avoid orange-and-black butterflies. But the viceroy isn't what it appears to be: despite its garish garb, it's a perfectly edible insect. It spends its formative instars feeding not on poisonous milkweeds but rather on innocuous willows, poplars, and other trees in the family Salicaceae. The viceroy thus gains protection from predators by pretending to be something it's not.

Actually, it's not just the adult viceroy that is guilty of false advertising; every life stage of the viceroy is a bit of a sham. The viceroy begins life as a spherical light green egg laid at the tip of a leaf, where it looks remarkably like the various and sundry pale green spherical galls that afflict the leaves of willows and poplars. Out of the egg hatches a bizarre-looking caterpillar, olive green or brownish, with a white blotch on its back, a row of knobby protuberances along its length, and a spiny antennalike projection on its thorax. What with its overall shape and coloration, a viceroy caterpillar more closely resembles a bird dropping than a caterpillar—a guise sure to discourage any and all birds with discerning palates. The caterpillars pass the winter in a folded leaf, pretending to be invisible, and pupate in early spring.

The entire larval stage of the viceroy, then, is geared around maintaining a low profile. The adult, however, is about as conspicuous as insects get. By mimicking the monarch—or "representing the king," as a viceroy is wont to do—*Limenitis archippus* gets to mate, lay eggs, flit about, suck nectar, and engage in other butterfly pursuits in comparative peace.

PAINTED LADIES

If you've ever seen a rosy-tinted brown butterfly with black spots on its hindwings and white spots on the tip of its forewings around town, you have plenty of company. The painted lady butterfly, *Vanessa cardui*, has been seen around towns throughout North America, Europe, Australia, New Zealand, Asia, and Africa. It's known on every continent except Antarctica, and it's even found on islands in the middle of the Pacific Ocean. In North America the butterfly ranges from sea to shining sea, as far north as the Pacific shores of British Columbia and the Atlantic shores of Labrador (and there is at least one report that it summers happily at the Bering Strait). Inland, it's equally undaunted by cold weather: it frequently flies at elevations of 7,000 to 8,000 feet, and one Dr. Packard collected specimens on Arapahoe Peak at 12,000 feet and others within 500 feet of the summit of Pike's Peak. Not surprisingly, the painted lady is known in some circles as the cosmopolite.

What is surprising, though, is how little is known about the biology of this example of one of the world's most common insects. One possible reason it attracts so little attention is that it rarely infringes upon human activity. In most places it spends the winter as an adult, taking shelter under leaf litter or in other unglamorous places. Around mid-

Painted ladies

May the butterflies quit their winter quarters and begin to lay eggs. The adult females are particularly enamored of plants in the Compositae, or daisy family, and prefer to lay eggs on thistles (earning them the sobriquet, "thistle butterflies"). The eggs are also found on cocklebur, burdock, yarrow, pussytoes, and other plants in the family.

This affection for the thistle family makes the growing caterpillars heroes in Montana and the plains states, where the painted lady keeps the weeds under control in pastureland. In California, however, they're regarded as pests for occasionally straying onto thistlelike crop plants such as artichokes and safflower. They will, if pressed, feed on plants in other families, particularly the Malvaceae, where they make themselves unwelcome on hollyhocks, mallow, and cotton.

The spiny caterpillars are pink with yellow longitudinal stripes and abundant black spots. They feed by pulling together leaves, buds, and leaf fragments with silk threads and weaving the whole thing into an irregular little camp site. The caterpillars often equip the upper half of their abode with thistle spines clipped off the leaves and woven into the silk webbing. Younger caterpillars merely scrape the upper cuticle of leaves, leaving the lower cuticle intact; larger caterpillars consume entire leaves and they eventually eat themselves out of house and home. They construct increasingly larger and more irregular nests until they spin a vaguely defined cocoon and pupate. One to two weeks later, the butterflies emerge and start in all over again until winter sets in.

The painted lady owes its cosmopolitan distribution to its own immoderate habits. Populations in some places build up to such a point that caterpillars are often left without a crumb to sustain themselves. The adults then undertake massive migratory flights. These flights can take the painted lady from California to the Canadian Rockies, a distance of over fourteen hundred miles. They've been clocked at eight mph on these long hauls.

When not on the road, the painted lady butterflies are more erratic fliers, frequenting fields, highways, and gardens. They have the unusual habit of picking out a place to bask and returning to it if disturbed. As one nineteenth-century admirer, Samuel Scudder, wrote in 1889, "It loves to return to the spot from which it has been driven, or to the immediate vicinity, often circling about first, as if selecting the best spot. Scarcely observing the pursuer, heedless of the net, it returns directly to the place it has left and sits with horizontally opened wings. . . . It is a nimble, lively, youthful untamed petulant insect." Maybe that's what it takes to make it as a world-class butterfly.

PARSNIP WEBWORMS

One of summer's roadside attractions, aside from sweetcorn or lemonade stands, is *Pastinaca sativa*, the wild parsnip. The parsnip has never been well loved as a vegetable, and it's even less likeable when it escapes from cultivation and establishes itself along roadsides, railroad tracks, and in abandoned lots or farm fields as a weed. Among other things, in sunny weather the plant can cause an intensely painful blistery rash with the slightest touch, a habit not likely to endear it to casual and unsuspecting passersby. But the wild parsnip has one admirer that has literally gone to the ends of the earth (or at least the Western Hemisphere) in hot pursuit. This insect afficionado is aptly named the parsnip webworm, *Depressaria pastinacella*.

When the Pilgrims landed at Plymouth Rock neither parsnip nor the parsnip webworm were anywhere to be found. However, when the colonists first began to set up shop, they (for reasons known only to themselves) opted to bring over a few seeds from Europe and begin cultivating parsnip in the New World. It took the parsnip webworm a little longer to colonize, since it did so without the aid or cooperation of Pilgrims. It was first reported on parsnips in Ontario, Canada, in 1865.

Parsnip webworm

Unlike people, the parsnip webworm eats not the root of the parsnip plant but rather the developing flowers and seeds. Every spring, just as the yellow umbrella-shaped flowers begin to make an appearance, so do the caterpillars. They're distinctively cream-colored with small black spots; the larger caterpillars have a yellowish cast to their undersides. At first they confine their feeding to buds and flowers along one spoke of the umbrella-shaped inflorescence. As they grow, they eat more and more florets, reeling them in with silk and webbing the whole affair into a cozy little nest. They are then free to feed in the privacy of their own home. If the infestation is a heavy one, virtually all the flowers on an umbel can be consumed by the time the caterpillars are ready to pupate.

Unaccustomed as they are to public display, the webworms burrow into the stem to pupate, presumably free from prying eyes or probing bird bills. After ten to fourteen days, a greyish brown, squarish-shouldered, rather nondescript moth about three-fifths of an inch in length emerges. The habits and hangouts of the adult moths are on the mysterious side. The operating assumption is that they spend summers and fall sipping nectar on occasion but are otherwise largely unoccupied. In late fall they seek shelter in protected places, such as under bark, leaf litter, or in abandoned buildings, and appear again in spring to mate and lay eggs on their much-loved parsnip plants.

The parsnip webworm eats very little else besides parsnips, and very little else besides the parsnip webworm eats parsnip plants. One principal reason is that the chemicals in the foliage and flower of the parsnip that cause the blistery rash in sunny weather are also extremely toxic to insects. Only a few insect species as a result have developed a taste or tolerance for the plant. One other parsnip devotee is *Agonopterix clemensella*, a close relative of the parsnip webworm. Unlike its relative, *A. clemensella* feeds only on the leaves, rolling them into tight tubes and seldom venturing out in the light of day. Inasmuch as sunlight is required to activate the insecticidal chemicals in parsnip, you could say that *A. clemensella* has it "made in the shade."

PRAYING MANTIS

Not long ago, when the New York state legislature was empowered to designate a state insect, the legislators selected three as candidates: the honey bee, the ladybug, and the praying mantis. Their choices were entomologically ill-advised to say the least. The honey bee was introduced from Europe (and hence is not exactly a native New Yorker); the ladybug is actually a complex of four hundred species in North America; and the praying mantis has some personal quirks that make it less than suitable as a state representative.

Mantids answer just about everybody's description of a "beneficial" insect. There are six hundred species worldwide that belong to the family Mantidae, and just about every single species is a voracious predator of insects. Of the two species most common in eastern North America, the Chinese mantis, *Tenodera aridifolia sinensis*, was intro-

Praying mantis: "Let us prey . . ."

duced into North America in 1869 in the hopes that it would ride roughshod over the local insect pests. The Chinese mantis is hard to miss; from head to tail it measures over three inches in length. The other species, the Carolina mantis, *Stagmomantis carolina*, is a native American and measures a more conservative 2½ inches in length at maturity.

As in all mantids, the front legs of the Chinese mantid are greatly enlarged and characteristically folded as for prayer—hence the name "praying mantis." In actuality, "preying" mantis is probably more appropriate, since the enlarged spiny forelegs are used not to salvage the souls of fellow arthopods, but instead to grasp them in a viselike death grip while the mantis leisurely consumes them. Most mantids are grey, green, or brown and blend in with the background; they characteristically perch on vegetation and wait for some inattentive victim to blunder by looking for pollen, nectar, or foliage to feed on. The mantid then lunges forward and locks in its victim. The move is so swift and deadly that it has given rise to an entire style of kung fu (known not inappropriately as "praying mantis style") in the Orient.

The personal life of the mantis is not exactly worthy of emulation. In late fall, females lay from fifty to three hundred eggs in a brown foamy egg capsule, called an ootheca, attached to vegetation. The eggs overwinter—the outer foam acting as an insulating layer—and develop in early spring. When the eggs hatch, the tiny immature mantids are ravenously hungry and immediately set about the business of consuming each other. The mantids in general are highly cannibalistic, and the first meal of a typical mantis is likely to be a sibling. People who are interested in raising mantids in captivity and collect oothecae in a jar are advised to separate the young ones immediately upon hatching unless they're content with one very large mantis per egg case. Development to maturity is a slow process and depends to a great extent upon the availability of insect prey. Mantids are not terribly particular and have been observed to feed on beetles, bugs, grasshoppers, bees, and even butterflies and moths.

Mantis courtship and mating are unique even among insects. When observed in the laboratory, sexually mature males stalk females through the vegetation. When they get close, they leap onto the female's back. The female then leisurely reaches over her shoulder and begins to chew off the male's head. Once the head is removed, the male's abdomen begins a series of movements that bring the genitalia into contact, and mating is complete. A male that is captured and whose head is eaten before it can mount is still capable (*sans* head) of walking around the female, climbing on her back, and mating. In nature, as opposed to the laboratory, it's likely that male mantids are more successful in escaping intact after their job is done. However, some entomologists have maintained that certain male mantids are incapable of mating while their head is still intact. The brain constantly sends inhibitory messages to the thorax and abdominal ganglia and it is only when the inhibition is removed along with the head that mating behavior takes place. This sort of inhibition is one that psychology provides no solutions for. In the interest of equal time, though, it must be said that decapitated females in the laboratory were observed to mate even faster than usual, and so they must also shed a few inhibitions along with their heads.

Mantids are undeniably important as natural control agents in the regulation of insect pest populations, but at the same time they regularly practice homicide, parricide, and a number of other "cides" as well. Such antisocial behavior, by human standards, is perhaps not so surprising in the insect world, where mantids are the closest living relatives of the cockroaches—another group that engages in behavior humans generally don't approve of. At least they don't make a regular practice of eating their mates.

SULFUR BUTTERFLIES

If you've ever wondered why they call them "butterflies," take a look at the next *Colias* butterfly you see; the whole subfamily Coliadinae of the family Pieridae is full of bright yellow or orange butterflies that resemble nothing so much as pure cholesterol on the wing. Their yellow color has also earned them the name "sulfur butterfly."

The two most common species in North America by far and away are the clouded (or common) sulfur, *Colias philodice*, and the orange (or alfalfa) sulfur, *Colias eurytheme*. The two are very similar in appearance and life cycle. Both species of caterpillars are green with lateral white lines (which may contain some red color in the orange sulfur). Both feed on plants in the legume family, including clover, trefoil, alfalfa, and vetch, and both overwinter primarily in the pupal stage. The easiest way to tell them apart as adults is that, if the wings of one contain any amount of orange at all, it's fairly safe to call it an orange sulfur. Both species, however, are basically yellow butterflies with black borders on the upper wing surface. To make things even more confusing, both species are prone to albinism or loss of pigment. In parts of their ranges there are sulfur butterflies that are white rather than yellow (in Alaska, for example, up to 95 percent of the clouded sulfurs are white). To add to the confusion even more, both species vary in the size and extent of the black wing border, particularly on the undersides of the hind wing. Needless to say, entomologists have

Sulfur butterflies: "I never know what to put on my name tag at these things, either."

had a hard time recognizing these species over the years, and, as one early entomologist lamented, the orange sulfur has been given more names than any crowned head of Europe.

Lest the entomologists get depressed too quickly over their failings, it should be pointed out that even the sulfur butterflies have a hard time sorting themselves out. Interspecific hybridization, or mating between species, is common throughout their range. Hybridization between different species is generally not a good idea, and the sulfurs have developed a safeguard against it, namely, an elaborate courtship ritual.

Courting and mating are major occupations of the sulfur butterflies, and it's probably no coincidence that the name *Colias* comes from a Greek epithet for Venus, the goddess of love. Far from shy about the whole thing, they lean toward the afternoon delight, carrying on their business primarily over the lunch hour (about 11 A.M. to 1 P.M.). Males patrol an area searching for any light-colored object that could conceivably be a female sulfur butterfly.

If a virgin female in the air or on the ground is ready, able, and willing, the whole thing is over fairly quickly. The male approaches and buffets the female with his wings, but the buffeting is not spouse abuse. By rubbing his wings together, the male activates specialized scales on his upper hind wing surface (called sex brands) that release an aphrodisiac chemical signal. A willing female responds by fluttering her wings, and, with that, courtship ends about 1½ seconds later (although actual copulation can last over an hour).

It's a different story if the female has already been mated or is just not interested. When a male approaches, instead of remaining stationary an unwilling female takes off straight up into the air. The male picks up in hot pursuit, and the two spiral around each other going higher and higher until the male finally gets the message and heads off elsewhere in search of yellower pastures.

Females can travel up to sixty feet in the air on these ascending flights before a persistent male literally gives "up." The females use several cues to determine beforehand if a particular sulfur butterfly is suitable material for mating. Among other things, the wings of the orange sulfur reflect ultraviolet light (which is visible to butterflies and invisible to us), whereas the wings of the common sulfur do not. So when it comes time for a female orange sulfur to choose a mate, "UV or not UV," *that's* the question.

WOOLLYBEARS

The woollybear caterpillar is a familiar sight in many places that are unfamiliar to most insects—just about everybody's seen them tooling down the road looking for all the world as if they're late for an appointment. This peripatetic habit is by no means of recent origin. In 1608, Topsell, in his *History of Serpents*, described "another sort of these caterpillars who have no certain place of abode, nor yet cannot tell where to find their foode but, like unto superstitious Pilgrims, do wander and stray hither and thither, consume and eat up that which is none of their own. . . . They never stay in one place, but are ever wandering [and] . . . by reason of their roughness and ruggedness some call them Bear-worms."

There are at least 120 species of woollybears, or arctiid caterpillars, in North America, but by far the most familiar is the reddish brown-and-black, neatly clipped, banded woollybear, *Isia isabella*, the woollybear of common parlance. It is unusual among the members of its family in that it spends the winter not in the dormant pupal phase but as a caterpillar. Every fall, the caterpillars seek out a suitable hiding place in which to spend the winter—hence, the sight of woollybears busily making their way across roads, along hedgerows, and in general joining the flow of traffic. In the spring, they repeat the trip, this time searching out food after a long winter's abstinence.

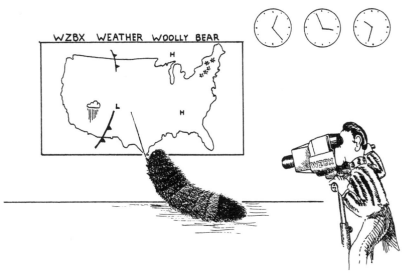

Woollybear

Topsell aptly described the woollybears' dietary habits as well: "They can by no means endure to be dieted and to feed upon *certaine* herbs and flowers, but boldly and disorderly creep over all and taste of plants and trees indifferently and live as they list." To say the woollybear isn't fussy is something of an understatement. The caterpillars are known to feed on over thirty plants in twenty-one families; their diet includes everything from elm trees to sweet clover. Not surprisingly, woollybears at times make their way into home gardens and wreak havoc, riddling the leaves of many garden plants in late summer and fall. A simple preventative is an aluminum foil barrier four inches high set around the margin of the garden. While they're great at walking, all those legs make them less than adept at climbing.

An adult woollybear, by the way, is called a tiger moth, because many members of the family sport black-and-orange stripes. *Isia isabella*, however, has yellow wings about two inches across with a few black spots here and there. Females lay their eggs in spring in patches on the leaves, and the caterpillars reach full size in four to eight weeks. The next spring the caterpillars weave a cocoon (incorporating their abundant supply of hairs), pupate, and emerge as adults a week or two later.

The banded woollybear isn't the only one to hit the road each fall. Yellow woollybears, *Diacrisia virginica*, are the yellow ones that, compared to the banded woollybear, look as if they need a haircut. *Estigmene acrea*, the salt marsh caterpillar, is greyish or black with a black head and frequently marches en masse, moving from field to field in phalanx formation. All the species of woollybears are about equally omnivorous and collectively can be found on more than one hundred species of plants.

The banded woollybear is perhaps best known for its meteorological capabilities. It's a long-standing piece of folk wisdom that the width of their reddish brown band predicts the winter weather—the broader the band, the more severe the upcoming winter. The scientific status of the legend is still hotly contested. C. Howard Curran of the eminently respectable American Museum of Naural History in New York correlated band width with weather conditions between 1947 and 1951 and actually found that the caterpillars predicted weather more accurately than some professional meteorologists. Curran used this finding as evidence of the predictive powers of the woollybear; it could also have been a demonstration of the current state of the meteorological art.

Chapter 6

Bush Leaguers and Fruit Pickers

Apple maggot flies

Black vine weevils

Cecropia moths

Eastern tent caterpillars

Grape phylloxera

Gypsy moths

Hackberry psyllids

Plum curculios

Tiger swallowtails

APPLE MAGGOT FLIES

Although most people look forward to autumn as harvest time for crisp, juicy apples, *Rhagoletis pomonella*, the apple maggot, gets a head start on eating apples by at least a few months. The apple maggot fly, a member of the Tephritidae, or picture-winged flies, is one of the most serious pests of apples in the northern parts of the U.S., ranging west to the Dakotas, south to Texas and Florida, and north to Nova Scotia, Canada.

Rhagoletis pomonella spends its winters innocuously enough, buried six inches deep in soil, encased in a hardened capsule, or puparium, about a quarter of an inch in length. When summer begins, in late June, the adult flies emerge. As flies go, they're fairly striking in appearance, with shiny black bodies, yellowish legs, and clear wings

Apple maggot flies: "She's from a hawthorn. We wanted to get married but we just couldn't settle on a date."

boldly patterned with black bands. After a week or so of scoping out the territory, females mate and begin the business of reproduction. Female flies search for developing apple fruits, preferably early varieties such as Cortland or Delicious, with sweet or subacid flesh. Once she finds a suitable fruit, a female punctures the apple skin with her sharp pointed ovipositor and inserts a single tiny white egg in each puncture wound. After ovipositing to her heart's content in a fruit, the female then drags her ovipositor along the fruit's surface tracing out large circular paths; as she drags she deposits a chemical signal, or phero-mone, which marks the fruit as hers. Subsequent female flies landing on the fruit, detecting the pheromone, fly off in search of fresher un-preempted forage.

With the marking pheromone, a female *Rhagoletis pomonella* guar-antees that her maggot children will have ample room in their apple to tunnel around and feed. The characteristic sign of apple maggot damage is the presence of brownish irregular tunnels through the apple flesh. This sort of damage has earned *R. pomonella* the appel-lation (as it were) of "railroad worm." Early in the course of their development apple maggots often leave no external signs of damage, but as fruit ripens the tunnels show up as dark lines under the skin. In a heavy infestation an apple is soon converted to a brown pulpy mass teeming with whitish, legless, pointy-headed maggots. Early-ripening varieties of apples are most suitable for growth and thus are most prone to damage. In an early variety maggots can complete development in as little as two weeks, particularly if the apple drops to the ground in the middle of it all. In late fall or winter apples, the best the apple maggot can do is to cause corky streaks; maggot development in these apples can take three months or more.

Name notwithstanding, apple maggots didn't always eat apples. In fact, apple maggots were around in North America long before apple trees were brought over from Europe by the early settlers. Originally, apple maggots are believed to have infested the miniature applelike fruit of *Crataegus*, or hawthorns, a member of the same plant family as the apple. A variety (or race) of *R. pomonella* still infests hawthorns throughout the range of the apple maggot. Why the insect made the switch is easy to see; growing up in an apple as opposed to a hawthorn is like growing up in a thirty-two room mansion instead of an efficiency apartment. *How* they made the switch is still subject to speculation. Entomologists now believe that it may all have been a matter of good timing. Hawthorn flies emerging early in the summer after hibernation would be forced to begin searching for fruits to lay eggs in before hawthorn fruits were available (they mature in August and September). In June and July, the search for round red fruits led the flies directly

116

to apples. The earlier the flies came out to oviposit, the sooner their offspring completed development, reducing the probability that apple flies would ever even encounter the seasonally distant hawthorn flies in their day-to-day activities. The end result is that there are two separate strains of *Rhagoletis pomonella* that have little to do with one another. Simply stated, it was all a matter of the early *worm* getting the apple.

BLACK VINE WEEVILS

If someone takes your vine before its time, a likely suspect is *Otiorhynchus sulcatus*, the black vine weevil. The black vine weevil is anything but a high-profile insect. The adults, black snub-nosed weevils about a fourth- to a half-inch long, with patches of golden yellow hairs on their sides, spend most of their summer daylight hours hiding around the roots of strawberries, blackberries, blueberries, raspberries, loganberries, and grapes, as well as several types of shrubs. Males are so secretive that there is some debate (perhaps even among female black vine weevils) as to whether they exist at all. At nightfall, females, whose wings are fused together, of necessity walk in and about the berry plants chewing scallop-shaped chunks out of foliage. In late June or July, the females add some variety to their life and begin to lay eggs (with or without benefit of mating) around the crowns of the vines or

Black vine weevils: "I've found yew at last!"

118

on the ground nearby. One black vine weevil can lay up to 1,600 eggs in her lifetime on the vines and can alternate egglaying with eating for up to a month.

The eggs hatch in about ten days, and the immature grubs—pinkish white, C-shaped, legless affairs with a brown head capsule—keep a low profile just like their mother. Underground, the larvae have their way with roots, often causing considerable economic damage. Larval feeding goes on through midsummer and fall and picks up again in the spring after the grubs pass the winter in diapause. Pupation occurs in early spring, and adults emerge by June to hit the berry patches.

Actually, the black vine weevil has plenty of company in the berry patch. The strawberry root weevil, for one, *Brachyrhinus ovatus*, is also known as the strawberry crown girdler. *B. ovatus* is, like the black vine weevil, shiny black with golden hairs. It's also flightless, and it's also almost entirely female in gender. The strawberry root weevil is an even more serious pest of berries than the black vine weevil; so also, for that matter, is the rough strawberry root weevil—*O. rugosostriatus* in the West and *O. rugifrons* in the East. These guys lack the golden trim and are slightly smaller than the black vine weevil, rarely exceeding a quarter-inch in length. They also tend to overwinter as adults under leaf litter, when they have to, and inside houses, when they can gain access.

In the last analysis, the black vine weevil actually does more damage to shrubs than it does to berries, particularly to broad-leaved evergreens such as yew and hemlock and to ornamentals such as rhododendron. The larvae consume feeder roots and the bark on major roots and can even kill nursery stock, particularly after it's transplanted. Yew bushes in particular are a great favorite of *Otiorhynchus sulcatus*, which is sometimes known as the taxus weevil (after *Taxus*, the Latin name for the yew). It's virtually the only weevil that feeds on yew, and the damage it does is easily recognized—notched and severed needles near the main stems of the plant.

Recognizing the damage is one thing, but stopping the damage is something else entirely. The black vine weevil was once easily controlled by drenching plants with a residual insecticide—in the soil to kill developing grubs and on foliage to stop the adults. Since the sixties, however, black vine weevil has shown resistance to most of the standard insecticides. An alternative approach is to rotate crops so that susceptible species are not planted in an area where beetles have become established. Rotating nursery tree species in areas where black vine weevils are known to exist, then, is just one way to know they'll never find another yew. . . .

CECROPIA MOTHS

Every year people say that the groundhog comes out of hibernation on February 2 to check the weather; if it sees its shadow, then it heads back underground for six more weeks of hibernation. The cecropia moth, *Hyalophora cecropia*, does the groundhog one better. It hedges its bets every year automatically without counting on a weather report.

Basically, like many North American moths, the cecropia moth passes the winter in hibernation (or diapause), its development arrested at the pupal stage, wrapped snugly in a large silken cocoon. Since *H. cecropia* is the largest moth in North America, with a wingspan of four to five inches, its two-inch cocoons are conspicuous in bare trees in the wintertime, strapped securely between parallel stems close to the ground or snugly ensconced among dead leaves in the upper branches.

Cecropia moth

After a long chilling period, development begins again in order for the adult moth to emerge in the spring. Ten weeks at 42°F (6°C) or less is enough of a chill for cecropia moths—for some, but not for all, that is. In any population of cecropia, there are two types of individuals: those who emerge in early spring and those who emerge when the weather is much warmer, almost two months later. The difference is genetic and results in two separate periods of emergence in the spring, a nifty way for the moth to avoid being wiped out altogether by a late spring frost.

Cecropia cocoons can be found in almost any tree or shrub, because cecropia caterpillars eat the leaves of almost any kind of tree or shrub—at least twenty-six species in over twelve families have been reported, including willows, birch, elm, hawthorn, black cherry, maple, dogwood, ash, and lilac. Moths emerge in the spring, usually in the morning a few hours after sunrise, and spend the day expanding and drying their impressive wings. They're hard not to recognize, with reddish brown wings marked with a red crescent on the front wing, a white cresent on the hind wing, and a white stripe and clay-colored border along both wings.

Just after sunset the males take off and fly in search of females, who obligingly remain in one place, exposing glands that produce a long-distance attractant, or pheromone. Males can fly seven miles or more in hot pursuit. Once mated, the female begins to lay eggs (about three hundred in five to six days) on the foliage of trees and shrubs. Since neither the male nor female moth eat at all, they rarely live very long. Their offspring, the cecropia caterpillar, in contrast, spends much of the summer eating. The bluish green caterpillar, with a row of bizarre yellow protuberances along its back, two rows of blue bumps on the side, and two red bumps on the tail, reaches about four inches in length before it spins a cocoon and pupates.

Cecropia cocoons are familiar to many people not only because they're so easy to spot in winter, but also because you don't have to travel far to see them. In the Midwest, population densities of cecropia are over ten times greater in town than in the woods or in the countryside. This is true even though the same species of trees can be found in both places. Evidently, cecropia thrive in town along tree-lined avenues and in front lawns and such because their major enemies (thrashers, woodpeckers, cardinals, and field mice), who can spot the cocoons in the bare branches as easily as humans, are considerably less abundant there than they are out in the countryside. *Hyalophora cecropia*, then, bucks the trend and comes to the *city* to "get away from it all."

EASTERN TENT CATERPILLARS

For almost thirteen years the Reverend Jonathan Fisher of Blue Hill, Maine, labored to produce a book entitled, *Scripture Animals, or Natural History of the Living Creatures Named in the Bible*. The book, published in 1834, was at best only a modest success; of the 1,000 printed, the Reverend himself purchased 625 copies. In the book, Reverend Fisher combined natural history lessons with biblical expostulation. Under "Caterpillar," after a brief but thorough account of scriptural references (in both Latin and Hebrew), he described,

the Caterpillar of New England . . . which infests the apple-trees in the middle parts of New England . . . has a black head and tail, its body . . . divided into ten segments or joints. . . . Each segment has a brush of yellow bristles, rising erect from the back of it; the middle part of the back is light umber, on each side of it is a yellow stripe. The side of each section is umber, having, towards the belly, a black, oval spot, in the center of which is a round spot of white. A blue stripe runs along on each side of the belly, and there is a narrow, blue ring, round the body between each segment.

Eastern tent caterpillars: "Don't worry, Gretel, I left a trail of silk. We'll find our way home."

Reverend Fisher, citing 1 Kings 8.37, drew a lesson from the caterpillar and its ilk—"when devouring insects greatly multiply in a land, it is for the sin of the people of that land." The people of New England must still be pretty sinful, since *Malacosoma americanum*, the "Caterpillar" described by Fisher, is still multiplying and devouring apace.

The life history of *Malacosoma americanum* has remained largely unchanged since the days of Reverend Fisher. Adults (chocolate-colored moths with yellow or whitish bars across the wings, which span just under two inches) emerge from their pupal cases in autumn. Their principal concern after emergence is mating and egg laying. Since they have no functional mouthparts, they can't feed and thus live only a short time. The female moths lay up to twenty-five hundred eggs in tight clusters coated with a waterproof foamy brown substance, and *M. americanum* spends approximately nine months of the year, including the winter, in the egg stage.

In May, when the weather turns warm, the caterpillars hatch and begin feeding on the tender young leaves of the tree in which the egg cluster was deposited—usually cherry, apple, or other species in the family Rosaceae. Soon after feeding, the caterpillars construct a silken web, or "tent." As Fisher describes, "They . . . weave themselves a silken nest, ten or twelve inches in diameter, upon some branch of the tree. Here they repose by night; by day they wander over the tree, eating the leaves of one branch and then of another, pulp, veins and all, to the stem." They molt six times over a four to six week period, and while naive young caterpillars content themselves with plants in the Rosaceae, including cherry, apple, pear, and other fruit trees, the more worldly and more ravenous older instars leave the trees on which they hatched to feed on anything within crawling distance, including maples, birch, chestnut, dogwood, beech, ash, witch hazel, walnut, sweetgum, blackgum, alder, oak, black locust, willow, basswood, elm, and blueberry.

As tent caterpillars travel among the branches in search of food, they lay down single strands of silk and follow these trails back to the tent after they finish feeding. Reverend Fisher suggested that they may orient themselves by following these trails and even conducted an unclerical experiment in which he rubbed the silk trails with his finger and watched the caterpillars "seem for a little while to be bewildered, groping for their way like a blind man." More recent research has confirmed that the caterpillars do lay down a chemical signal along the silk they lay as they travel. Not only can they find food by following the trails of their nestmates, they can find out how extensive the menu will be when they get there; well-fed caterpillars lay more attractive trails than underfed caterpillars do.

Trailblazing is not entirely without risks, from the perspective of a tent caterpillar. Since nestmates, and even different species of tree-feeding insects (for example, *Malacosoma disstria*, the forest tent caterpillar), can follow trails of *Malacosoma americanum*, it stands to reason that other insects, including insects that eat tent caterpillars, can do the same thing. Indeed, some predaceous stinkbugs make their living tracking down tent caterpillar trails and leisurely poking and sucking dry caterpillars when they reach the mother lode; one species actually lives inside the tent and captures tent caterpillars as they return from their feeding trips. So living in a silken web with your nestmates, surrounded by chemical signals clearly pointing the way for potential enemies to find you, is not a peaceful carefree way to live. The Reverend Fisher did not comment on whether tent caterpillar problems with "devouring insects" result from *their* sinful ways, however.

GRAPE PHYLLOXERA

Bastille Day, July 14, commemorates a historic occasion in France when the bourgeoisie rose up against the privileged classes for their repeated violations of the basic principles of liberty, equality, and fraternity. However, history takes little note of an even greater crisis in French history that arose a century later. In 1860, an aphid was accidentally introduced into France that singlehandedly (or six-leggedly) almost destroyed that bastion of French culture, the wine industry—namely, *Phylloxera (Dakutisphaira) vitifoliae*, the grape phylloxera. This native of North America can be found today wherever grapes are grown.

There are at least four versions of grape phylloxera: its life cycle wins no awards for simplicity, and to some extent it still keeps entomologists guessing. Grape phylloxera spend the winter either as eggs on grape stems or as yellow wingless aphids, less than one-twentieth of an inch long, dormant in galls on the grapevine roots. In spring, the eggs on the stems hatch, and the newly hatched aphids move to the new leaves to set up residence in leaf galls; those overwintering in roots simply

Grape phylloxera: "It's a full-bodied but flavorful vintage, a little playful on the stylets."

get in gear as spring arrives. To make things even more complicated, the leaf gall–formers at maturity give birth asexually to living young that either infest more leaves or drop to the ground to join the root gall aphids already resident. In the fall, the root gall–formers produce winged individuals that climb up to the soil surface, leave the ground, and lay eggs on the vines. But, as they say in late-night television commercials, "Wait, there's more"—these eggs hatch into males and females which mate, and the mated females then lay one egg apiece on the grape cane to spend the winter.

To make things even more confusing, the different forms of the grape phylloxera are found with different frequency in different parts of the United States. Surprisingly, the East Coast phylloxera are more laid back than their West Coast equivalents. On the East Coast, the leaf gall form of phylloxera predominates and rarely causes much damage, but in the West the root gall form predominates and is so destructive it can stunt or even kill vines in three to ten years.

What does all of this have to do with the price of wine in Bordeaux? In the middle of the nineteenth century, infested American grapevines were imported into Europe, and the famed French vines, completely unfamiliar with phylloxera, succumbed with alarming frequency. The resulting encroachment across Europe has been called "the greatest single viticulture disaster since the flood." Within a quarter of a century, almost a third of all French grapevines (extending over 2½ million acres) had been destroyed.

French ingenuity is the reason that fine French wine (post-1895 vintage) is still available. French enologists realized that the American fox grape, *Vitis labrusca*, was unusually resistant to the deadly root gall form of phylloxera. They then imported American rootstocks and grafted stems of the French wine grape *Vitis vinifera* onto the hardy American roots. At an overall cost of about ten billion francs, the phylloxera problem was solved and the wine industry saved. The solution, however, may only be temporary. California vinters who imported non-resistant French rootstocks are currently facing their own phylloxera outbreak, and there is some evidence that a new strain of phylloxera has established itself in California that can infest even resistant rootstocks. In France, however, the situation appears to be holding—a testament to one of the most ingenious conquests of an insect invasion.

GYPSY MOTHS

An otherwise forgettable astronomer by the name of Leopold Trouvelot made entomological history of a sort in Medford, Massachusetts, in 1868. An amateur entomologist, Trouvelot was interested in improving upon methods of silk production and imported several species of silk-producing caterpillars from Europe. For reasons perhaps known only to himself, Trouvelot quickly abandoned both his project and his caterpillars; one species, *Lymantria dispar*, quietly slipped out of his yard and into neighboring woodlands. Thus the gypsy moth was introduced into North America, and the name of Trouvelot (at least in certain circles) lives in infamy.

At first glance, the name "gypsy" seems to be a misnomer; the white-winged females are so heavy-bodied that they can barely fly. However, the smaller brown males can and do fly and tend to fly in an erratic zigzag pattern heading nowhere in particular. In mid- to late July, the females emerge from cocoons, and, after mating, the process of egg-laying begins in earnest. Eggs are laid in masses of one hundred to one thousand and are covered with a fine coat of buff-colored hairs.

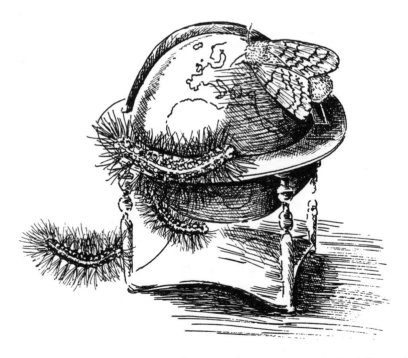

Gypsy moth and caterpillars: "Tell us how you came over on the boat, Gramps!"

The eggs are plastered on objects anywhere within reach of the less than agile female—mostly on tree trunks, under stones, on buildings and the like. The eggs remain quiescent all winter and hatch the following spring. After a few days of nibbling, the tiny caterpillars—the real gypsies—send out thin silken strands, catch a wind current, and start traveling. The hair on their bodies is hollow, increasing their buoyancy. They can reach heights of two thousand feet and travel for impressive distances. Ballooning, for example, is most likely how the gypsy moth first got across Cape Cod Bay in Massachusetts.

The caterpillars eventually settle on a tree and, not being terribly particular, begin to eat. They've been recorded on over five hundred species of trees although, admittedly, they have their favorites, among them oak, apple, willow, beech, aspen, birch, and basswood. The caterpillars are voracious eaters and rapidly increase their size from less than one-fourth of an inch to two inches in length, stripping trees of foliage in the process. When full-grown, the caterpillars are easy to recognize—pale brown with stiff brown and yellow hairs projecting at all angles, bedecked with five pairs of blue spots and six pairs of red spots along the back. By late June, the caterpillars spin a cocoon and pupate. They emerge as moths in two weeks and begin the cycle all over again.

The gypsy moth has more than lived up to its name. The first outbreak occurred in Medford, Massachusetts, and quickly spread to the neighboring state of Connecticut. By 1924, the federal government created a "barrier zone," a region thirty miles in width running from Long Island north to Canada. The idea was to keep the states west of the line free of gypsy moths. Needless to say, the quarantine was a disaster, and gypsy moths are now at home as far north as Maine, westward to Minnesota and Texas, and south to Maryland and Delaware. Occasional outbreaks are reported in midwestern states.

While ballooning is certainly an effective form of dispersal, it pales into insignificance in contrast with the travel assistance rendered to the moth by humans. The gypsy moth's propensity to lay eggs on low-lying objects has worked greatly to its advantage. Such low-lying objects include lawn furniture, plants, logs, campers, and mobile homes that travel regularly with their owners from New England to far-flung places.

The government still keeps the gypsy moth under quarantine—nurserymen must have their stock inspected and certified as gypsy-moth–free before shipping, and massive monitoring programs are being carried out in all states where the gypsy moth is a threat. Monitoring, it turns out, is a simple affair. The flightless female, to enhance her chances of finding a mate, produces a sex attractant, or pheromone, which draws in males from all directions. Scientists have synthesized the pheromone, dubbing it disparlure (from the old name *Porthetria dispar*), and bait sticky traps with it; they then sit back and wait to count all the hopeful males that enter the traps looking for females and fun.

One enterprising scientist at Michigan State, in an all-out effort to keep tabs on gypsy moths, even trained German Shepherds to track down and sniff out egg masses, providing a graphic example of how of entomologists are "doggedly" pursuing this pestiferous insect.

HACKBERRY PSYLLIDS

Hackberries (*Celtis occidentalis* and its close friends and relatives) are widely planted as windbreaks and shadetrees throughout the Midwest, and wherever they occur they almost invariably develop an assortment of bumps, protuberances, and pimples on their leaves, buds, and stems. These abnormal growths are galls formed by hackberry psyllids, insects that can be found wherever hackberries are found.

Psyllids are tiny soft-bodied plant-feeding insects that resemble dollhouse versions of cicadas without all the noise. Actually, they're not too distantly related to cicadas, which, like the psyllids, have piercing-sucking mouthparts and uniformly membranous wings (as do all members of the order Homoptera). Psyllids are distinct within the order in that they come with stout, thick hind legs, which they use to make miniature leaps and bounds around their host plants (earning them the name, or expletive, jumping plant lice).

Hackberry psyllids all belong to the genus *Pachypsylla*, a group ranging in size from barely a quarter-inch to under one-eighth of an inch. The insects, however, compensate for their small size with remarkable powers of increase. *Pachypsylla celtidisvesicula* makes blisterlike galls about one-sixth of an inch in diameter that cover upper leaf surfaces. *Pachypsylla celtidismama* makes so-called nipple galls, which rise to a height of a quarter-inch or so all over the undersides of leaves.

These two species pass their days in more or less the same manner. Adults of both species come out in spring and wait for hackberry leaves to bud out. Eggs are deposited on the leaves and hatch in seven to ten days. After the nymphs begin to feed, the plant tissue immediately

Hackberry psyllid

around the feeding site begins to grow in an abnormal and distorted manner, eventually giving rise to a gall (wherein the soft-bodied, otherwise vulnerable insect can feed in peace). They take most of the summer months to complete development and emerge in September and October as adults, whose first priority is to find a safe warm spot to spend the winter. At this point, hackberry gall makers make their presence known to the community at large by hurling themselves on and adhering to window screens, screen doors, cars, fresh paint, and lawn furniture around houses. They're basically harmless (if annoying), not only to the people whose homes are invaded but also to the hackberry trees, which appear to sustain no permanent damage from being infested.

Hackberry bud gall psyllids, *Pachypsylla celtidisgemma*, are less conspicuous than their relatives, the leaf gall makers. Unlike the other psyllids, the bud gall psyllid spends the winter inside the gall as a nymph and emerges in June the following year. The adults lay eggs in June as flower buds form. The newly hatched nymphs then move from leaf to bud and form galls in the developing flowers. *Pachypsylla venusta*, the hackberry twig gall maker, forms a large spherical gall up to two-fifths of an inch across on hackberry twigs.

Snugly ensconced in their galls, the hackberry gall makers are protected from most casual predaceous passersby. However, there are several chalcid wasps which make it their profession to search out and consume psyllids in their galls. These wasps, aptly named *Psyllaephagus* ("psyllid eaters") and not so aptly named *Torymus*, use their ovipositors to pierce the gall and lay eggs inside. Once hatched, the wasp grubs make short work of the psyllids. One study showed that up to 30 percent of the bud gall makers and 50 percent of the nipple gall makers are destroyed every season by chalcid wasps. So sometimes, then, even hackberry insects just can't hack it.

PLUM CURCULIOS

Sir Issac Newton was purported to have observed applies falling in an orchard one day, and his training and insight led him to suggest that gravity was the force behind the fall. If Newton had been an entomologist, however, gravity might never have been discovered. Entomologists more or less concur that apples fall, at least in June and July before they ripen, due to the work of the plum curculio, *Conotrachelus nenuphar.*

The plum curculio is a small blue-black weevil with a penchant for plums, peaches, pears, apples, and other fruits in the family Rosaceae. Adult weevils bide their time in winter in orchards under leaves, rocks,

Plum curculios: "Direct hit!"

and fences or in nearby woodlands. They become active just about the time apple trees bloom and feed on the developing flowers. After they mate, the females eat a small round hole into the skin of the fruit, using an enormously elongated snout built expressly for this purpose. The female then turns around, deposits an egg in the freshly made cavity, and turns around again to cut a crescent-shaped scar directly below the egg. An enthusiastic female can lay up to one thousand eggs in this laborious manner, although the average is closer to two hundred. After a while, the exposed flesh in the scar starts to decay, and the grubs hatching from the eggs two to twelve days later feast on the rotting flesh for the duration of their development.

The immature plum curculio is not much to look at. It's a greyish white C-shaped grub lacking legs altogether and almost lacking a head (one has to look closely to see the tiny brown bump at the anterior end). After hatching, the grub gradually gorges its way through the fruit to the central seed cavity. After about three weeks it eats its way out again and drops to the soil, where it builds a small pupation chamber. Depending on the temperature, an adult weevil emerges in a month or so to nibble on fruits, pitting and scarring as it goes, until cold weather puts a damper on its activities.

Now, the sight of apples falling to the ground is hardly newsworthy, but, in an orchard infested with plum curculio, fruits fall not when they're supposed to, in August or September, but prematurely, before they ripen—so early that frustrated farmers refer to them as June drops. A normal fruit falls because the tree secretes an enzyme called pectinase, which breaks down the cementlike substance, pectin, that holds plant cell layers together. Once the layers become unglued, the fruit is free to fall. The plum curculio short-circuits the entire process by producing its own pectinase, fooling the tree into shedding its fruits prematurely.

The plum curculio hit upon this bit of biochemical wizardry not merely to impress its friends and win admirers, but to ensure its own survival—living in a ripening fruit is a hazardous occupation. During the process of ripening, sap and expanding fruit cells in the growing apple close off passages tunneled out by the larvae, eventually killing them by asphyxiation or by physically crushing them. Once it's off the tree, however, a fruit stands little chance of ripening, and the weevil is free to tunnel hither and yon.

While the grub has a stake in making fruits fall, farmers have a stake in keeping them on the tree, as free from blemish as possible. Thus, plum curculios have not endeared themselves to fruit growers. Besides the losses brought about by early fruit fall, many fruits are economically unmarketable due to unsightly pitting and scarring. Moreover, ovipositing females can spread brown rot of peach and plum, a severe fungal disease. One of the most effective control measures is sanitation—cultivating the soil in late spring to kill larvae and pupae in the ground and collecting the drops in early summer and destroying them to interrupt the life cycle. Once plum curculio is taken care of, apples can then fall when they're ripe—just as nature (and Newton) intended.

TIGER SWALLOWTAILS

Unbeknownst to most people, there are tigers lurking in backyards across most of eastern North America. These tigers, however, aren't likely to inspire much fear or do much harm to anyone. The tiger swallowtail, *Papilio glaucus*, is so named because it sports yellow wings with black stripes. As butterflies go, it's hard to miss; with a wing span of up to four inches, it's one of the largest butterflies in the country. The name "swallowtail" refers to the long tail-like extensions of the hind wings, the hallmark of the family Papilionidae. The butterfly has four black stripes on each yellow forewing and a prominent V on each hind wing. They're frequently seen visiting blossoms of clover, lilac, thistle, or bee balm.

The tiger swallowtail is striped for much the same reason that a tiger is striped; the stripes break up the outline of the body, so it is difficult to discern in the vegetation. A peculiar thing, however, happens throughout part of the range of the tiger swallowtail: from Florida and Texas north to North Carolina through Illinois, the female tiger swal-

Tiger swallowtail

135

lowtail loses her protective stripes. Up to 95 percent of the females in some parts of the range are jet-black, with a margin of yellow spots and an iridescent blue patch on the hindwing—more of a "black panther" than a "tiger." This geographic change in color has puzzled entomologists for decades. Samuel Scudder remarked in 1889 that "to explain these phenomena is exceedingly difficult. The moment one begins to speculate some fact appears which altogether upsets any theory he may form on the subject." The general consensus at the moment is that the black color is also protective, like the stripes. Female black tiger swallowtails bear an uncanny resemblance to another swallowtail, *Battus philenor*, the pipevine swallowtail. As caterpillars, pipevine swallowtails feed on Dutchman's pipe and other plants in the genus *Aristolochia*. These plants contain aristolochic acids, extremely noxious chemicals poisonous to mammals, birds, and even to most insects. The pipevine swallowtail stores these chemicals in its body (as a caterpillar) and then in its wings (as an adult). Birds learn very quickly that black butterflies taste bad and as a result avoid black butterflies in general.

The tiger swallowtails, irrespective of color, are perfectly edible, having grown up as caterpillars on such innocuous species as birch, tulip, poplar, ash, wild cherry, and magnolia. But wherever black tigers occur in the range of the pipevine swallowtail, they escape bird predation by virtue of their resemblance to their distasteful relations. Indeed, tiger swallowtails aren't the only ones in the family capitalizing on the resemblance. The black swallowtail, *Papilio polyxenes*, and the spicebush, or green-clouded, swallowtail, *Papilio troilus*, are also perfectly edible, but are black with a row of yellow spots on the front wing margins and an iridescent blue patch on the hind wings.

The tiger swallowtail caterpillar is not above engaging in a little deceit as well. When it hatches from its greenish egg, it is black with a white "saddle" mark across its back and is a dead ringer for a bird dropping. Since no self-respecting bird would eat a bird dropping, the caterpillar is protected by the ruse. As it gets bigger, the guise wears thin, and after its third molt, the caterpillar turns green, with two prominent eyespots behind its head. At this point, it is a good stand-in for a tree snake. It even has a bright orange, forked gland behind its head to flick in and out like a snake's tongue.

The name "tiger swallowtail" seems quite suitable, but the name *Papilio glaucus* is not so obviously apt. Carolus Linnaeus, the eighteenth-century Swedish scientist responsible for the system of naming animals and plants in use today, decided for reasons known only to himself to name the swallowtails after Greek heroes of the Trojan war. Glaucus was a Lycian prince and ally of Troy. Upon meeting Diomedes, a Greek, in battle, Glaucus exchanged his golden armor for the brass armor of the Greek as a gesture in recognition of the long-standing friendship of their families. The act has since become immortalized as a symbol of an unequal exchange—but perhaps Linnaeus was unaware of the fact that, when the tiger swallowtail exchanges its bright stripes for a drab black exterior, it's getting the best of the deal.

Chapter 7

Underground Gourmets

Antlions

Cicada killers

Ground beetles

Patent leather beetles (bess beetles)

Periodical cicadas

Rock crawlers (grylloblattids)

Sowbugs

Springtails

Winter stoneflies

ANTLIONS

Antlions, or insects in the family Myrmeleontidae (from *myrme*, for "ant," and *leo*, for "lion"), inspire no doubt as much fear and loathing in the ant world as their mammalian counterparts do in the furry warm-blooded world. *Adult* antlions are rather unremarkable. With their long slender abdomen, membranous wings crisscrossed by a fine network of veins (typical of all members of the order Neuroptera), and their knobby (or clubbed) antennae, they are, if anything, dainty and delicate—hardly horrifying. The immature myrmeleontid, however, is the stuff of which ant nightmares are made. The larvae, sometimes called doodlebugs, have a wide, flattened abdomen and a constricted thorax topped with a head equipped with strong sickle-shaped mandibles as long as the head and thorax put together.

These distinctive jaws serve several functions. Antlion larvae, unlike lions, don't waste their energy frantically stalking prey and running it down. Instead, they set a trap. Using its head and elongated mandibles as a shovel, an antlion, walking backward, digs a cone-shaped pit in sandy soil. It then retreats to the bottom of the pit, opens its mandibles, and waits. When an unsuspecting ant happens by, it slips down the side of the cone-shaped pit and, unable to regain its footing in the sandy soil, slips inexorably closer to the open mandibles. If it reaches the jaws in proper position, it is, according to W. M. Wheeler's 1935 account, "instantly seized, buried, paralyzed and sucked out; but if it

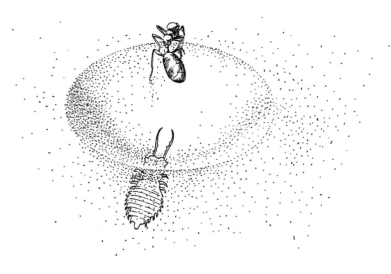

Antlion: "Another unexpected pit stop."

is at first awkwardly or inconveniently grasped, it is either repeatedly tossed into the air or against the walls of the pit till the proper hold has been secured, or if it escapes is showered with sand till it is again brought within reach. If this fails, the antlion may actually leave its pit and pursue the prey, though this procedure, owing to the insect's retrograde mode of locomotion, is not apt to be very successful." An individual antlion makes several pits during the course of its life; after consuming the requisite number of ants, it spins a silken cocoon in which to pupate (itself not a mean feat, given that sand sticks to every strand of silk).

While it is true that not all myrmeleontids make pits (some simply conceal themselves under loose debris), it's also true that not all pit-makers are antlions. Another denizen of the dust that strikes fear into the dorsal aorta of ants is the wormlion, the larva of rhagionid flies in the genus *Vermileo* (*Vermi* for "worm," *leo* again for "lion"). That *Vermileo* constructs pits is remarkable, since it labors under several major handicaps; namely, as a maggot, it has no appendages to speak of, and its head is so far withdrawn into its body as to be useless as a digging tool. The maggot digs its pits, usually in sandy areas near cliffs, by crawling on its back, inserting its anterior end backwards and downward into the sand and rhythmically contracting its body to toss out sand in a spray. Once at the base of its pit, it remains on its back waiting to seize a hapless victim and inject it with a potent paralytic salivary toxin. Since it lies on its back, a wormlion literally has eyes in the back of its head for enhanced ant watching. The main problem faced by ant and wormlions is that it may be a long time between meals. Setting traps means long periods of waiting for an ant to wander by. Some might even say that, as a mode of predation, it's the pits.

CICADA KILLERS

One of the most familiar sounds of summer is the lazy droning trill of the dogday cicada, or harvestfly. Nobody looks forward more to hearing it than *Sphecius speciosus*, the giant cicada killer. *Sphecius speciosus* is, as enormous wasps go, a very attractive insect. One of America's larger species, it tops an inch and a quarter in length, and its basic black body, with yellow racing stripes on the abdomen, and shiny yellow wings make it hard to miss. For their own needs, cicada killers content themselves with sucking nectar from summer flowers. But when it comes to looking after their children, cicada killers live up to their name.

Every year in early August female cicada killers busy themselves digging a tunnel in a dry spot in a bank or roadside. The tunnels average about six inches in depth, then turn right angles for another six or so inches, and terminate in large circular chambers. The wasp accomplishes this feat by judicious use of her feet; walking backward on four back legs, she uses her two forelegs to carry dirt along the burrow and out the entrance hole. Once the burrow is complete, *Sphecius speciosus* sets out to provision it, or stock the shelves, so to speak. To this end, she rounds up cicadas.

Cicada killers

It's not clear how cicada killers find their prey, but it almost certainly isn't by sound—the wasps are equally as likely to capture female cicadas as males and, unlike males, female cicadas don't sing. Once a cicada is sighted, the wasp moves swiftly, grasps the victim, and stings it, injecting in the process one of the fastest-acting paralytic toxins known. If you've ever heard a cicada trill abruptly turn into a slur and then silence, now you know why. The venom doesn't actually kill the cicada, it just keeps it paralyzed. Moving the cicada (which can weigh up to six times as much as the wasp) back to the burrow is hard enough without having to manipulate a conscious struggling victim.

If the wasp catches a cicada on or near the ground, she can't take off and fly, so instead she drags the cicada on its back up a tree trunk, as though it were a sled, and launches herself into the air, falling back downward along an angle. She repeats the process as many times as is necessary to get the cicada back to her burrow. Once there, she attaches a single white egg onto the cicada's body, usually on the underside near the front legs, and deposits it in one of the hollowed-out chambers. In a few days, the wasp grub hatches and placidly begins to consume the still-living, and thus farm-fresh, cicada thoughtfully provided by its mother. After the cicada is consumed completely, the larva spins a cocoon and remains in its chamber underground through the winter. Come summer, it pupates, emerges, and gnaws its way out of its subterranean chamber to reach the surface—much, no doubt, to the chagrin of all the cicadas busily emerging from their own underground chambers at roughly the same time.

Although *Sphecius speciosus* is very particular as to what it eats, it's not stupid. While it's true that the only form of food it provides to its offspring is cicadas, it doesn't take just any cicada. *Sphecius speciosus* never feeds on periodical cicadas, the ones that only appear every 13 or 17 years. Instead, it concentrates its attention on the annual cicadas, species that complete development in a year or two and emerge every summer. Thus *Sphecius speciosus* may have actually pioneered the concept of "fast food."

GROUND BEETLES

Unlike groundhogs, when adult ground beetles exit their underground mud cells, they probably never see their shadow, no matter what the time of year they emerge. Among other things, ground beetles, species in the family Carabidae, are almost exclusively nocturnal. Their preference for the dark of night is one possible explanation for their lack of publicity and accounts at least partly for the sad fact that a family with over twenty thousand species worldwide and almost twenty-five hundred species in North America alone can walk along a busy street and not be recognized by fans. Moreover, the majority of ground beetles don't attract a lot of attention even if someone does cross their path; they tend toward dark conservative browns to blacks in overall body color.

One probable exception among the ground beetles in the dress-for-success category are the species in the genus *Calosoma*. The name *Ca-*

Ground beetles

losoma translates roughly from the Greek into "beautiful body" and is descriptive, to say the least. Species in the genus are generally brilliantly colored, often even iridescent. Like most other ground beetles, *Calosoma* species are voracious predators of other insects, particularly caterpillars. *Calosoma* are also exceptional among ground beetles in that they're among the handful of species that, name notwithstanding, actually climb up into the trees when they've exhausted the caterpillar supply at ground level. Some of the smaller species have modified hair pads on their legs to assist them in their lumberjack number.

Among the more spectacular species in the group is *Calosoma sycophanta*, the gypsy moth hunter. *C. sycophanta* was imported from Europe in an effort to bring down the numbers of another European import, *Lymantria dispar*, the gypsy moth caterpillar. *C. sycophanta*, ranging from approximately ¾ to 1¼ inches in length, is almost impossible to miss, not only due to its size but also due to its flashy colors. Its abdomen is brilliant brassy green, and its thorax is a dark metallic blue. The adults, which can live for several years, spend the winter in underground chambers. In spring, adults emerge and lay eggs in mud cells in the soil. The immature ground beetles, glossy black elongated spindle-shaped grubs, are predaceous like their parents and are partial to gypsy moth caterpillars. Within the two-to-four-week period of its development, a single grub can consume over a dozen gypsy moth catepillars, often climbing trees (like its parents) in pursuit of comestibles.

Although *C. sycophanta* is a European import, there are some closely related homegrown varieties of ground beetle as well. *C. calidum*, the fiery hunter, is a plain black beetle about an inch in length. It's strictly earthbound and never climbs trees in search of its favorite foods, cutworms and armyworms. The fiery searcher, *C. scrutator*, ranges throughout the United States and southern Canada. This beetle can compete with the foreign imports for style. Almost 1½ inches in length, it comes in black with a violaceous sheen on its abdomen; its wing covers are iridescent green, lined with purple; its legs and thorax are steel blue; and the legs of the males are equipped with red tufts of hair. Even the larvae are on the gaudy side, yellow in color with a brown head and brown abdominal plates. While *C. scrutator* may not win any awards for subtlety, it's certainly one ground beetle not content to remain just a "background" beetle.

PATENT LEATHER BEETLES (BESS BEETLES)

John, Paul, George, and Ringo may have been Beatles with a unique sound, but *Odontotaenius disjunctus* and its relatives in the family Passalidae are "beetles" with a unique sound. The Passalidae is a group that goes under a number of names. They're also called peg beetles, horn beetles, bess bugs, betsy beetles, or patent leather beetles. The latter name reflects their shiny, shoe-leather black coloration. There's only one species of passalid in North America—*Odontotaenius disjunctus*, which has itself been known by a few other names, including *Passalus cornutus* and *Popilius disjunctus*.

The passalids are relatively unusal among beetles in that they are capable of making a remarkable variety of sounds. *Odontotaenius disjunctus*, known to possess a fourteen-signal repertoire, is not only the most vocal among insects, it produces a greater variety of acoustical signals than any fish, amphibian, or reptile, and greater than that of many birds (and at least a few punk rock bands). Adult bess beetles produce sounds by rubbing the dorsal portion of their fifth abdominal segment, which is studded in rasplike fashion with tiny spines, against the sclerotized, or hardened, folded portions of their membranous hind wings. The part of the wing in contact with the rasp on the abdomen is equipped with transverse ridges lined with backward-pointing spines, like a file. The whole process is known as stridulation (from a Latin word meaning, "to make a harsh sound").

Unlike crickets, cicadas, and katydids, passalids are affirmative action stridulators; both sexes are capable of sound production. Passalid sounds can be broken down into pulses, bars, and phonatomes. A pulse is basically a single sound unit that can be produced by a single strike

Patent leather beetles: "She bugs you, yeah, yeah, yeah . . ."

of the rasp against the file; a bar is a series of pulses separated by a silent interval; and a phonatome is a complex sound cycle. These pulses, bars, and phonatones combine to form signals that are appropriate to different social situations. The Type A signal—a long bar or phonatome—occurs when the beetle is annoyed, threatened, or otherwise disturbed, either by a predator or by another beetle—unless the beetle happens to be a member of the opposite sex. Courtship in the Passalidae is a noisy process. Males initiate the process with what might be considered a singles "bar." Females respond with a characteristic sound, and signals bounce back and forth for up to twelve hours until mating takes place. The actual process of mating, which lasts only ten to thirty minutes, is the only silent part of the whole process, since in many species males produce a postcopulatory signal as well (although other species are discreet enough not to kiss and tell). Passalid mating habits are the source of considerable conversation among entomologists for reasons other than acoustical. Among other things, they are the only beetles known to copulate face to face in missionary position.

For entirely unrelated reasons, passalids are considered to be among the most socially advanced of beetles. They live in colonies in rotting logs. The adults feed premasticated wood chips to their grubs and assiduously tend the pupae to protect them from predators. Passalid larvae can also produce sounds. Their third pair of legs is reduced in size and worthless as locomotory equipment, but each leg serves as a plectrum which rubs against a file on the upper joint on the second pair of legs; they take the words right out of their feet, so to speak (or so to stridulate). The function of sound production in the larvae isn't quite clear, but it's thought to function in maintaining colony integrity and in soliciting parental care; passalid beetles obviously don't subscribe to the human philosophy that children should be seen and not heard.

PERIODICAL CICADAS

For millions of people across the country, New Year's Day is cause for celebration, an occasion for ushering in the New Year. For a periodical cicada, however, January 1 is yet another in a seemingly endless series of uneventful days, New Year's or otherwise.

The periodical cicada is one of the more familiar insect prodigies. Everybody's at least passingly acquainted with its life story. After it emerges from an egg laid under bark in May through July, the young cicada drops to the ground, settles into the soil, makes itself a little nest, and plugs into a nearby tree root. Typically, the cicada doesn't see the light of day for the next seventeen years, passing the time engaged in the same routine—sitting and sucking. Then, as if by pre-arranged signal, all the cicadas in a neighborhood emerge at once, climb up tree trunks, shed their skin, and make an incredible racket. In an outbreak year, cicada populations can be so dense that they can clog swimming pool filters and stop traffic as cars skid on the bodies of the many misdirected individuals searching for a trunk to conquer. Up to forty thousand can emerge from under a single large tree.

Periodical cicadas: "Long time no see . . . say, you wouldn't happen to know who won the last sixteen World Series, would you?"

149

The din is entirely the responsibility of the male; it's his way of announcing his arrival (after seventeen years, no less) to female cicadas. This sex difference has been known for centuries and in fact prompted one chauvinistic wit, Xenophon, in classical times to remark, "Happy are the cicadas / for they all have voiceless wives." The noise is produced by the action of muscles vibrating a pair of drumlike organs in the cicada's thorax, or midsection. An air cavity acts as a resonator and connects to the outside through a pair of tiny holes. The resonant trill thereby created is an effective form of communication for the male (a way of getting things off his chest, one might say).

Not surprisingly, cicadas have called a considerable amount of attention to themselves. Their conspicuousness has given rise to numerous misconceptions as well. First of all, strictly speaking (or rather, entomologically speaking), the seventeen-year locust is not a locust at all. Locusts are the leathery-winged, grasshopperlike creatures that were responsible for one of the biblical plagues. The cicada actually belongs to the order Homoptera (the sucking insects) and is merely called a locust due to its populations of plaguelike proportions. Moreover, not all cicadas wait for seventeen years to emerge. There are six species of periodical cicadas in North America, and only three species—*Magicicada septendecim*, *Magicicada cassini*, and *Magicicada septendecula*—operate on seventeen-year cycles. The other three species—*Magicicada tredecim*, *Magicicada tredecassini*, and *Magicicada tredecula*—emerge from underground after a mere thirteen years.

Moreover, the periodical cicada is sometimes confused with the common dog-day cicada, or harvestfly. The dog-day cicada is thought to complete development in a breathtaking two years. One of the easiest ways to distinguish between the periodical cicada and the dog-day cicada, short of waiting seventeen years to be absolutely certain, is to look for red eyes, a distinctive feature of the seventeen-year variety. One other distinctive feature of perodical cicadas is a wing marking resembling a black W, which in less enlightened times was thought to prophesy "war" or some other form of unpleasantness beginning with a W.

The periodical cicada's precise but prolonged time schedule has excited more than a little curiosity among entomologists, accustomed as they are to getting what they want in a hurry. The current theory to account for the thirteen- or seventeen-year cycles revolves around predation. When a large population of helpless juicy insects appears on the scene, predators generally make the most of the situation. Long-lived predators can remember the feast and return to the scene in subsequent years; short-lived predators, being well fed, reproduce successfully and exuberantly and leave a large population of predators to await next year's emergence. Next year's emergence, however, doesn't happen for another seventeen years, so the periodical cicada is able to outlast, and thus escape, enemies to a great extent. Which is not to say cicada populations aren't hard-hit during an emergence year. They're essentially defenseless—they can't run away or fight back, they produce no noxious defensive chemicals, and they don't even have the sense to give "playing dead" a shot. However, a singing cicada is just one voice in the crowd (and a large crowd to boot), and the mass chorus reduces the risk of capture of any individual cicada to a low level.

Surprising to most people is the fact that, despite their prodigious numbers, cicadas rarely do any serious damage to the trees on which they assiduously suck for so long. What injury there is can usually be blamed on the ovipositing females, who push the bark up from the wood in order to deposit from four hundred to six hundred eggs at a crack (as it were).

For residents of the eastern United States, seeing periodical cicadas is a matter of patience. In the West, however, there are no periodical species and no massive outbreaks. For admirers of insect accomplishment, a trip out east during an outbreak is certainly merited. After all, it's not just a good show, it's the "trill of a lifetime."

ROCK CRAWLERS (GRYLLOBLATTIDS)

While it's true that all insects are cold-blooded, some are colder-blooded than others—and grylloblattids, or rock crawlers, are without doubt the coldest-blooded of all. Grylloblattids are known from only three locations in the world, and all three are under snow cover for most of the year: the Rocky Mountains of the northwest United States and Canada, Siberia, and east Asia. Unlike us mammals, most insects can't regulate their body temperatures, so when the outside temperature falls much below 50°F, insects either migrate, go into winter diapause or hibernation, or get cold feet (which is a real problem if you have six of them). Grylloblattids are a conspicuous exception. They prefer temperatures around freezing and are still scurrying around at temperatures of 23°F (when most of their colleagues would be insect popsicles). If there's a temperature extreme grylloblattids can't take, it's heat. A comfortable 75° room temperature puts them into permanent paralysis, and the body heat from a human hand can kill them.

Although grylloblattids weren't discovered until E. M. Walker turned over some rocks at 6,500 feet in Baniff, Alberta, in 1914, they are undoubtedly very ancient insects. Entomologists are undecided as to precisely where they belong in the taxonomic scheme of things. Their name is a case in point. Grylloblattids resemble crickets (members of the family Gryllidae) in their flattened bodies and the shape of their heads, but they resemble cockroaches (members of the family Blattidae) in that their legs are equipped for roachlike running rather than cricketlike hopping. Resemblances don't end there, though. The female ovipositor is structurally very similar to the ovipositors of katydids or Tettigoniidae; the position of the antenna on the head is reminiscent of stoneflies or Plecoptera; and their overall appearance evokes earwigs or Dermaptera. As a result, grylloblattids are placed in their own order, the Grylloblattodea, where they will stay out of the way and confuse fewer people.

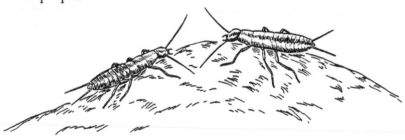

Rock crawlers: "Big deal! We made it—first grylloblattids on Everest. So who's gonna know?"

The life cycle and habits of grylloblattids are pretty much of a mystery inasmuch as most entomologists are reluctant to make field observations in subfreezing temperatures. They appear to be nocturnal and to prey on the insect odds and ends that share their frigid habitat. They may also occasionally nibble on fungi and mosses. They're most active in spring and fall and avoid the heat of summer by moving down into the soil to a depth of twenty inches or more. They forage actively in winter on the snow surface, traveling up to 360 feet in the course of a night's work, and can scuttle at speeds just under an inch per second.

No one really knows much about their life cycle. One guess is that over the course of development, the nymphs molt four times and reach maturity in a process that takes years. Even less is known of the grylloblattids' sex life. Observations have been made of captive specimens, however, and one event described by N. Ford in 1926 went as follows: "when the male and female were put together, they played their antennae a moment, then suddenly there was a rough and tumble fight and the male fell at the bottom of the cage. Slowly he came up the moss again, this time in a flash winding himself around the body of the female and catching the edges of her prothorax with his widespread mandibles. This hold he retained during a half-hour's struggling and tumbling until finally by a jujitsual method he made her powerless by slipping his first legs under her first and second ones and his second pair under her third. Copulation lasted 12 hours." For rock crawlers, then, "getting the cold shoulder" from a potential mate is just the thing to begin a beautiful friendship.

SOWBUGS

People are inclined to believe that any animal that's small, vaguely repulsive, and in possession of more than the usual number of legs must by rights be an insect. But sowbugs aren't any kind of bug at all. If you were pressed to come up with a next of kin, you'd be closer with barnacles than with bed bugs. Sowbugs (also known as pillbugs, woodlice, or slaters) are joint-legged arthropods that belong not to the class Insecta but rather to the class Crustacea. Their allegiance, then, is with barnacles, lobsters, crayfish, crabs, and water fleas, among others. The vast majority of crustaceans are aquatic, and the Isopoda, the order to which sowbugs belong, are no exception.

All of the landlocked isopods belong to the suborder Oniscoidea, and it was no mean feat for them to adjust to life on land. For one thing, sowbugs, like the rest of the crustaceans, breathe with gills, which must stay moist to operate, so on land the sowbugs are, for the most part, condemned to life in damp, dark corners. Also, raising off-spring on land has presented more than a few problems. Sowbugs have gotten around the need to return to the water to breed by pro-viding their eggs with a built-in swimming pool. Female isopods have a brood pouch, or marsupium—a water-filled cavity situated under-

Sowbugs: "I don't care if it is an essential mineral in my diet, I'm not eating it!"

neath the isopod thorax, or chest-equivalent. Fertilized eggs are cared for in the brood pouch, and the young isopods, after hatching, stay in the pouch until they're able to fend for themselves. The twenty-five to seventy-five youngsters look for the most part like miniature versions of their parents and need a year or more to reach full size.

Adults sowbugs have carried one-upmanship to extremes. Instead of three pairs of legs, as all self-respecting insects have, isopods have seven pairs, all identical in size and shape (hence *isopod*, Greek for "equal legs"). Instead of the usual insect single pair of antennae, isopods have two pairs, although the first pair is much reduced. Each body segment is topped with an armorlike plate that overlaps the plate on the section beneath.

The isopods that occasionally infest gardens, greenhouses, and basements are generally no more than a half-inch in length. *Porcellio laevis*, a sowbug proper, is brown, glossy, and smooth. *Armadillidum vulgare*, a pillbug, is slate grey in color to almost black. Both occasionally make a nuisance of themselves by nibbling the tender bases of plant stems close to the ground. The easiest way to distinguish between sowbugs and pillbugs is to disturb them. Sowbugs either run away or clamp down onto the surface of the ground with a fourteen-legged death grip that defies the efforts of most predators to dislodge it. Pressed against the ground, the dorsal plates touch the surface, making it very difficult even to pry them loose. Pillbugs owe their name to their habit of rolling up in a ball like armadillos when threatened (hence *Armadillidum*). Most predators, unable to gain a purchase, end up helplessly rolling their prey around for a while before losing interest (although a few simply pop them in their mouths and swallow them whole).

Sowbugs provide the ultimate example of recycling. Hard pressed to conserve water, they don't expel waste through water-based urine but instead release ammonia gas directly into the air. They're even capable of taking up water through their anus as well as their mouth. And they depend on copper to survive; it's a major element of their blood pigment, so they eat their feces to recycle it. So next time you complain about lugging aluminum cans to the recycling center, be grateful you don't have to recycle like sowbugs do.

SPRINGTAILS

There's nothing particularly vernal about springtails—the "spring" in the name doesn't refer to any seasonal preference at all. Actually, the spring in a springtail, any species in the class Collembola, is a structure called the furcula, which projects from the fourth or fifth abdominal segment. The furcula hooks into a knoblike "catch" on the third abdominal segment called the tentaculum. When the furcula is released from the tentaculum, it hits the ground with enough force to propel the collembolan up in the air, often for considerable distances—hence the name "springtail."

The furcula isn't the only unusual thing about collembolans. In fact, one would be hard pressed to find anything about them that's usual. They are so different from the usual run-of-the-mill insect that many people put them in a class of their own. Their body plan, either globular or elongate, with a tubelike device projecting from the first abdominal segment, is about 400 million years old, according to fossil finds. Unlike

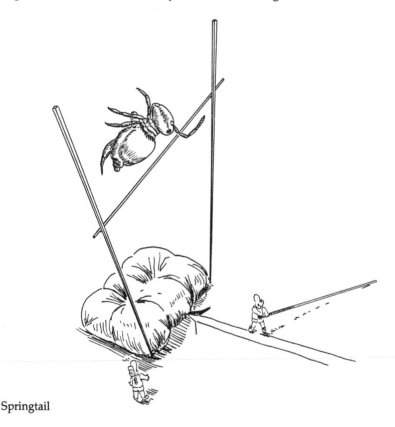

Springtail

most insects, collembolans lack wings—never had them and never will. Also, unlike most insects, collembolans lack the traditional air-tube tracheal system used for respiration. They apparently can acquire enough oxygen by breathing through their skin. Since oxygen moves very slowly though cuticle and can't penetrate much more than about a tenth of an inch, most collembolans aren't much wider than two-tenths of an inch across—a fact which doesn't make them among the more widely known and easily recognized arthropods. Finally, most self-respecting insects stop molting after they reach sexual maturity, but sexual maturity doesn't stop the collembolans. They continue to molt throughout their adult life, occasionally up to fifty times or more.

Their small size has made springtails difficult to study, but the studies that have been done suggest that their behavior is about as strange as their morphology. Springtails win no prizes for fine dining. In approximate order of frequency, they consume: fungal filaments, dead or rotting plants, insect droppings, pollen grains, spores, algae, bacteria, small animals (like rotifers, tardigrades, and nematodes), other collembolans, fungal juices, and regurgitated termite food. Their sex life isn't exactly a model of refinement either. Fertilization is accomplished by a technique called sperm packet transfer, the conceptual antecedent of the sperm bank. Basically, males produce a sperm drop, or spermatophore, and leave it lying on the ground. The exact details vary with the species; some will only deposit their sperm on the ground in the presence of a female, while others just casually leave theirs anywhere. Some go the extra mile and place their spermatophore on a long stalk, so their reproductive material isn't simply left lying on the ground. It's up to the female, then, to locate the sperm and essentially "sit on it"— pick it up and insert it into her genital opening. With all these sperm packets lying around on the ground, it's unclear how female collembolans distinguish spermatophores from males of the same species from inappropriate spermatophores. To make things even more complicated, males have a tendency to eat old spermatophores they happen across.

With such a haphazard mechanism for mating, one would think that there would be precious few collembolans around. Actually, nothing could be further from the truth. Although it's difficult to gauge population sizes, estimates of the number of springtails in an average cubic meter of soil range from 1,000 to 100 million (based on a sample from farmland). One reason behind their large numbers is their small size; approximately 13,000 to 15,000 individuals weigh one gram, which amounts to 405,000 springtails per ounce. The inevitable conclusion is that good things sometimes come in small packages—as any male springtail would tell you.

WINTER STONEFLIES

If you think that not having enough sense to come in out of the rain is the ultimate in stupidity, wait till you find out that there are insects without enough sense to come in out of the *snow*. Winter is usually the off-season for insects for several very good reasons, not the least of which is the fact that insects are poikilotherms—that is, they can't regulate their body temperature internally. So throughout much of the winter insects are literally cold-blooded, and at temperatures of 32°F or below most of them can't keep their metabolic machinery in gear. As a result, as winter sets in most insects either slow down or stop development altogether and enter a dormant period like hibernation called diapause.

There is, however, a perverse group of insects in the order Plecoptera (the stoneflies) whose development speeds up rather than slows down as temperatures drop. This is not an idiosyncrasy of one or two species; in at least four subfamilies of the family Nemouridae there are stoneflies that have undergone a calendar shift. In the subfamily Nemourinae, the spring stoneflies, adults appear from April to June. In the Leuctrinae, or rolled-wing stoneflies, adults appear as early as February in

Winter stonefly

the central states and December in the South. Not to be outdone, the Taeniopteryginae, or winter stoneflies, emerge in January and continue through April, and the Capniinae, or small winter stoneflies, emerge throughout any or all of the winter months.

All of the winter stoneflies are fully active during the winter months. They mate, they gambol on the snow, and they nibble on whatever greenery they can find, which means that it's mostly blue-green algae on the menu. Aside from their extraordinary temperature preferences, stoneflies are otherwise singularly unprepossessing. The adults are uniformly dull brown or black in color and more or less flattened in shape. About the only distinctive features they boast are a pair of many-segmented, tail-like appendages called cerci and a large lobe on the hind wings that is folded at rest. That's where the name of the order comes from—*Plecoptera* translates to mean "folded wing."

The name "stonefly" is more descriptive of the immature stages or nymphs—soft-bodied flattened creatures that tend to locate under stones in streams, rivers, or on the lakeshore. Like the adults, the nymphs have two long tail-like appendages and are otherwise pretty much undistinguished. They feed on anything ranging from plant material to leaf litter to other insects, depending on the species.

Just how winter stoneflies have made the adjustment to winter weather is not fully understood. While their dull brown or basic black color scheme won't win them any aesthetic prizes, it may be important in absorbing enough heat from winter sunlight to keep them going by solar heating. However they do it, though, they are rare among insects for having adopted as their credo, "there's snow place like home."

Chapter 8

Flexible Flyers

Carpenter and mason bees

Crickets

Dance flies

Fireflies

Hover flies

Katydids

March flies

Snowy tree crickets

Yellowjackets

CARPENTER AND MASON BEES

Regardless of the state of the economy, housing starts are up every spring for carpenter bees. The carpenter bees, a handful of species in the subfamily Xylocopinae of the family Anthophoridae, are known for constructing homes in solid wood. In the insect world, construction work is almost exclusively the province of the female. In the case of *Xylocopa virginica*, a black-and-yellow species resembling a large bald bumble bee, the female excavates a tunnel, usually parallel to the grain, into a tree or log (or occasionally, to the distress of homeowners, into sidings, shingles, eaves, windowsills, doors, fenceposts, telephone poles, or wood lawn furniture). These tunnels, of up to a foot or more in length, are subdivided into a series of rooms, with cemented wood chip discs serving as partitions. As the female remodels, she provides each chamber with a mass of pollen and nectar, deposits a single egg next to the mass, and then seals off the space to begin work on the next room. The hatching grubs feed on the pollen mass and eventually burrow out of their tract homes when they emerge into adulthood. There's only one generation each year; come winter, carpenter bees hibernate as prepupae or as adults in their cells and emerge early the

Carpenter and mason bees

following spring. Despite the resemblance to bumble bees, carpenter bees (which are glossier and larger) are loathe to sting, although if sufficiently provoked, the female is capable of inflicting some dandy pain.

The little carpenter bee, *Ceratina dupla*, approaches the home construction business in a manner that differs from its larger relative *Xylocopa*. These bees are metallic blue in color and are considerably smaller in size—barely reaching a quarter-inch in length, which is puny compared to *Xylocopa*, with its dimensions of an inch or greater. Instead of boring into tree trunks, they tunnel into the soft pith of dead twigs, especially sumac, syringa, blackberry, and other pithy plants. The female also partitions off a series of cells inside the hollowed-out stem and equips each cell with an egg and a mass of pollen and nectar. Unlike the larger carpenter bee, however, she moves in with her kids, occupying the uppermost cell while she waits for the family to mature. The lowermost larva is first to mature; it tunnels through the ceiling of its cell into the cell of its sibling, and the newly emerging bees successively make their way to the top. When everybody reaches the cell in which the mother is resting, mother and family all fly off together in search of nectar and pollen, which they collect on specialized hairs on their hind tibiae. Occasionally, one of the offspring will return to the old homestead, clear it out, and remodel it for her own use.

Carpenters aren't the only master craftsmen (or craftswomen) among the bees. Mason bees, in the family Megachilidae, also build homes in the stems of pithy plants. Like carpenter bees, they hollow out the pith and deposit an egg along with a mass of pollen and nectar in each chamber. Unlike the carpenter bees, however, mason bees wall off each chamber with a mixture of dirt or clay and plant material. Recognizing a mason bee is a bit difficult. *Hoplitis producta*, the most common mason bee in the eastern United States, is about a third of an inch in length and is black with bands of short white hairs along the sides of the abdomen. The taxonomic characteristics separating the family Megachilidae (the mason bees) from the Apidae (which includes, among others, the carpenter bees) are esoteric, to say the least, and involve such features as antennal sutures and mouthpart modifications. It's not surprising, though, that mason *bees* are hard to recognize; after all, Masons have long been known as a secret society.

CRICKETS

For the most part, insects appear in courtrooms only if somebody forgets to renew the contract with the exterminators; however, every now and then insects are actually the subject of litigation. Take the little-heralded court case of *Ben Hur Holding Company* v. *Fox* (1933). In that case, a judge ruled that an infestation of crickets was an insufficient justification for withholding rent money. In explaining his decision, the judge reasoned that, "While the cricket is technically an insect and a bug, it would appear from the study of his life, that instead of being obnoxious, he is an intellectual little fellow, with certain attainments of refinement, and an indefatigable musician par excellence." While correct on most counts, the judge made a few entomological gaffes. Among other things, a cricket is certainly not a "bug." A "bug" is technically a member of the order Hemiptera and is in possession of sucking mouthparts and half-membranous wings. A cricket, in contrast, is a member of the order Orthoptera and has the chewing mouthparts and straight leathery wings typical of the order. As for the "certain attainments of refinement," it's an odd choice of words to describe an insect that frequents garbage dumps and trash piles and is known to feed on everything from seeds, grain, and dead insects outdoors to paper, fur, and wool clothing indoors.

No doubt what wins over judges (and most other people) when it comes to crickets is their remarkable aptitude for singing. Unlike many animal sounds, the chirp of a cricket has the characteristics of a musical tone. The name "cricket" in fact reflects the human preoccupation with

Crickets: "Where's Buddy?"

the sound; it's from the French word, *criquer*, and means "little creaker." Crickets sing not with their throats (if insects can be said to have throats) but with their wings. The cricket forewing has a thickened vein crossing the wing near the base. This vein is etched like a file with fifty to three hundred microscopic ridges. On the upper side of the wing is a thick hardened area. When a cricket "sings," or stridulates, it lifts its wings over its body and pulls the file on the underside of one wing over the hardened "scraper" of the wing beneath it. The less leathery parts of the wings vibrate when the file teeth hit the scraper and amplify the sound.

Not all crickets are "indefatigable musicians." Among other things, nymphal or immature crickets lack wings and as a result don't have what it takes to stridulate. However, the adult females, who do have perfectly serviceable wings, never make a sound. Males generally sing to attract mates, and in fact the loudest and most common cricket song is the "calling song." Every species has its own unique calling song to ensure that females won't fall in with the wrong crowd. Females listen to the songs with the aid of a thin, flat membrane, the tympanum, on the lower part of their forelegs, in a spot that is roughly equivalent to a human elbow.

Female crickets respond preferentially to the call of their own kind, which is a good thing, because crickets are otherwise very hard to tell apart. What humans generally regard as *the* field cricket—the black or brown cricket, between a half an inch and an inch in length—is actually about six or seven species in the genus *Gryllus* that are almost indistinguishable until they start to sing. The field cricket oftentimes enters houses, where it can be confused easily with the so-called house cricket, *Acheta domesticus*. Both field and house crickets have adopted a similar life-style—egglaying in the ground in the spring, hatching in the summer, and molting eight to twelve times on the way to adulthood while feeding on plant leaves, dead insects, and clothing and carpet material, depending on whether they're indoors or out.

Over the years, people have come up with a number of unusual uses for crickets. In China, crickets are bred for their aggressiveness, and territorial fights, staged as sporting events, draw large crowds. Crickets are also raised commercially for use as pet food or live bait. One entrepreneurial individual in the late nineteenth century, a V. M. Holt, even went so far as to suggest that crickets may be particularly suitable food items for humans. To overcome any natural antipathy toward the idea, Holt offered a very compelling etymological argument. After all, as he phrased it, "their very name, *Gryllus*, is in itself an invitation to cook them."

DANCE FLIES

The spirit of Christmas is nowhere better exemplified in the insect world than among the dance flies. Dance flies, members of the family Empididae, are rather small unremarkable flies, rarely exceeding a quarter-inch or so in length. They're commonly found in moist places such as along watercourses or lakeshores. The larvae, also unremarkable, require much moisture and live in wet soil, in decaying wood, under bark, or under water. They look a lot like most other maggots except for aquatic forms, which have a series of well-developed false feet, or pseudopods, on most body segments. The name "dance fly" comes from the tendency of the adults in spring and early summer to form large aggregations suspended in mid-air prior to mating.

So far, nothing to deck the halls about. It's only when dance flies get down to the actual business of mating that they emulate St. Nick: virtually all dance flies give their mates a nice gift prior to copulation. The gift varies with the species. Empidids are almost all carnivorous and feed on other small flies such as mosquitoes, midges, and even on occasion other dance flies. One of the more common gifts presented to a female during courtship is a dead insect (it is, after all, the thought that counts). Since acceptance of the gift is a precondition for mating, some male dance flies actually wrap the prey with strands of silk and only then present the package, gift-wrapped, to the female.

There are some nontraditional gift givers among the family. In some species, the male consumes the juices from the gift item, wraps it up, and then presents the empty carcass to the female. Then there's *Hilara sartor* and a few other species. *H. sartor* males are vegetarians and feed only on nectar. Unlike dead flies, nectar is tough to wrap and transport, so a male *H. sartor* goes to great lengths to construct a delicate silk balloon as large as he is. He then carries his balloon to a swarm of other males and hangs around until a female comes by and picks him

Dance fly: "Empty again! That's it! Next time I eat him."

out of the crowd. He presents his balloon and, if he's lucky, she'll accept it and they'll go off to mate. The balloon is entirely empty and is essentially nothing more than an intricate spitball. As Harold Oldroyd describes in his *Natural History of Flies*, "it is as if one first presented a fiancée with elegant diamonds, then plied her with diamonds in an exquisite case, and finally offered her an elaborate but totally empty gift-wrapped box."

The behavior seems queer, to say the least. Prevailing opinion, however, is that the practice is not so much a reflection of gullibility on the part of the females but rather circumspection on the part of the males. Remember that the diet of most female dance flies includes other dance flies, so prospective suitors may be mistaken for prospective suppers. The "gift" may serve as a distraction to the female; while she's busy eating, the male can do his business and then quickly vacate the premises. Christmas among the dance flies has some rather different connotations; one old-time favorite may well be, "I Saw Mommy *Eating* Santa Claus."

FIREFLIES

Not all of the sparks flying on the Fourth of July are fireworks—more than likely a few of them are fireflies. Even among the entomologically unlettered, fireflies, or lightning bugs, need no introduction, but perhaps a few words of explanation would be in order. First of all, those insects with glowing posteriors that inevitably accompany the first days of summer are neither flies (as in fireflies) nor bugs (as in lightning bugs). They're beetles, and the term "firefly" refers to the two thousand members of the family Lampyridae, which translates loosely to mean "shining fire." The most outstanding feature of the family is the light-producing organ on the abdomen, or tail-section, of the adult beetle. The light produced by fireflies defies duplication by humans in that it's "cold" light—almost 100% of the energy produced by the chemical reaction responsible for it winds up as light energy. By comparison, the average electric light bulb loses about 90% of its energy as heat, with only 10% light energy produced.

What puts the "fire" in a firefly is a substance called luciferin. It's stored in cells of the light organs. These organs are richly supplied with air tubes. When oxygen comes in contact with luciferin in the presence of an enzyme called luciferase, it reacts chemically to release energy in the form of light. A layer of reflector cells in the light organs

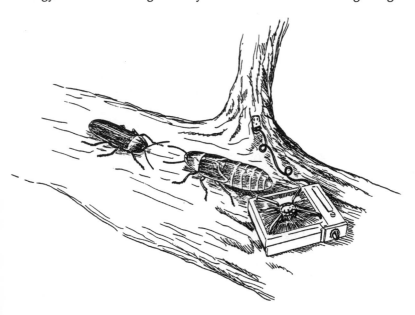

Fireflies: "Thank you, honey, for asking to have me for dinner tonight."

169

intensifies the effect. The light produced ranges from a yellowish green to reddish, depending on the species. The number and location of the light organs vary. Adult *Phryxothrix*, for example, have eleven pairs of green lights on the thorax and abdomen and a pair of red ones on the head—an arrangement worthy of a Broadway marquee.

Fireflies control light production by regulating the oxygen supply to the light organs. Every species flashes only intermittently, and the flash pattern is highly species-specific. The flashing is believed to help courtship and mate finding. Females generally become active around dusk and settle on some prominent piece of vegetation. Males cruise through the air anywhere from a few inches to a few feet off the ground, flashing their particular coded signal. Once a female spots the appropriate signal, she flashes back the same signal in response after a one- or two-second interval. After a few exchanges of five to ten reciprocal signals, the male homes in and lands, and they turn out the lights for the rest of the evening.

By and large, this is an extremely effective mate-finding system, but some fireflies have short-circuited the process. Females of the genus *Photuris*, like most female fireflies, climb up on vegetation and scan the skies looking for a signal. *Photuris*, however, can return the signal of just about any species of male firefly, not just members of her own species. It's not that they're promiscuous, exactly. After the appropriate number of exchanges, the wrong male lands next to the signaling *Photuris*, who promptly devours him. Instead of a mate, he's a meal—a flash in the pan, if you will. *Photuris* males are no doubt particularly careful not to get their signals crossed.

Most adult fireflies content themselves with small insects as dietary fare, and some don't feed at all. The immature larvae have elongated sickle-shaped jaws, which they use to inject a fast-acting paralytic toxin into slugs, snails, earthworms, and a variety of immature insects. The toxin also serves to liquify the body contents so the young fireflies can suck their prey dry. Immature fireflies, by the way, are sometimes called glowworms, as are the wingless females of some species, and, unlike the adults, glowworms produce light not at intervals but continuously. Even the eggs of some species are luminous. The function of light production in eggs and larvae is not well understood—perhaps research will someday "shed light" on the subject.

HOVER FLIES

There is no more easily recognizable herald of spring than the bee that takes to wing as the first blossoms of the season appear. But in all probability, the first "bee" of spring isn't a bee at all but instead is a syrphid or hover fly. Many flies in the family Syrphidae specialize in mimicking all types of bees and wasps. Harmless themselves, they assume the guise of a venomous insect to take advantage of the fact that birds and other predators tend to avoid pursuing and eating anything that even remotely resembles a poisonous and painful meal. The mimicry runs the gamut from sublime to ridiculous. Some flies sport only the black and yellow stripes characteristic of many wasps, while others mimic particular species down to the last tiny detail. The flies make up for the various anatomical obstacles to their masquerade in a number of ingenious ways. The leading edge of the forewing is darkened to resemble the wing folds of wasps, and darkened front legs dangle in front of the head to convert stubby fly antennae into long slender wasp antennae.

Hover flies: "I'm not eating that thing! Last year I ate one, and I was sick for a week!"

Like the bees and wasps they imitate, hover flies as adults feed primarily on floral nectar. In fact, they're probably second to bees in terms of their contributions to the pollination of economic plants. Their flight pattern, however, betrays them. While bees and wasps bob up and down when visiting flowers, hover flies, as the name implies, hover, or hang suspended in the air. When they land, they rest with their wings outstretched; bees and wasps fold their wings together. Finally, the big give-away: syrphids, like all "true flies," have only two wings, and all proper bees and wasps have four—a useful clue to anyone enterprising enough to get close enough to count.

Mimetic syrphids rely on the fact that birds and other enemies can remember painful experiences with bees and wasps and learn to avoid all bee- and wasp-like insects. To take full advantage of their enemies, many syrphids fly in early spring and late fall, whereas bees and wasps fly mostly in the summer. By flying in the spring, syrphids encounter birds that learned the previous summer to avoid bees and wasps; by *not* flying in the summer, they avoid the still-naive fledglings that might not yet have learned to avoid bees and wasps. So syrphids are abundant not when the stinging insects they resemble are common, but when experienced birds are common. So, if you're fooled by a hover fly some sunny spring day, don't feel bad—you've got lots of avian company.

The larvae of Syrphidae are a varied bunch. They are pink, green, or yellow legless blind maggots that crawl on vegetation and hunt aphids. When they find a colony, they pierce an aphid with their mouth hooks, hoist it in the air, and suck it dry. One species of syrphid can consume over one thousand aphids in only two to three weeks. Others live in bee and wasp nests, and some live underground and eat bulbs, roots, and fibers. A large number spend their formative days in gutters, drains, and sewers living in putrefaction.

Among these insect sewer rats is the rat-tailed maggot, *Eristalis tenax*, that lives in water so foul that its breathing holes or spiracles are located on a long extensible tube or siphon that telescopes up to the surface of the water to a distance of about five inches. The rest of the body *sans* "tail" measures less than one inch in length. Rat-tailed maggots are so enamored of sewage and excrement that from time to time they can't wait, and they go ahead and establish residence in the rectum of humans—a condition known euphemistically as intestinal myiasis. This condition is, mercifully, extremely rare in the more-developed areas of the world.

By and large, though, the syrphids are a good deal from the human perspective. They pollinate flowers, they destroy aphids, and, even though they resemble bees and wasps, they're sufficiently genteel to drop the resemblance when it comes to stinging or biting humans.

KATYDIDS

The nation's longest-lived controversy is still raging on just about every warm summer night, and the still-unresolved question as to whether Katy did or didn't is shouted into the darkness. Katydids, grasshoppers in the family Tettigoniidae, owe their common name to their very distinctive repetitive call. Like their close relatives the crickets, katydids make sounds by rubbing a raised ridge on one front wing against a filelike structure on the opposing front wing. There's one important distinction, however, between crickets and katydids—in katydids, the left wing folds over the right, with the ridge below the file, and in crickets, the right wing folds over the left, with the ridge above the file. Of course, the occasional wingless species of katydids must improvise. *Stenopalmatus longispina*, for example, makes noise (or stridulates) by rubbing pegs on its hind legs against projections on the side of its body—making it no doubt one of the original proponents of body language. While only the male katydid stridulates, both sexes can enjoy the racket—males and females are both equipped with "ears" on what roughly corresponds to their elbows.

Katydids are distinguished from other jumping, singing colleagues in the order Orthoptera by their extremely long antennae—thus, the appellation, "long-horned grasshopper." Unlike the short-horned

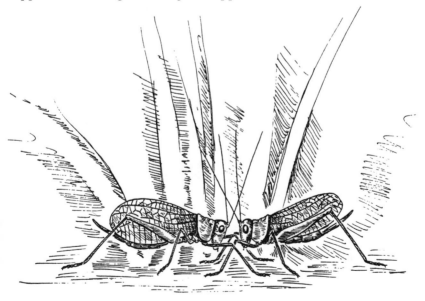

Katydids: "Did not!" "Did too!"

grasshoppers, katydids are by and large nocturnal and do all their singing at night. They're doubly difficult to track down in the dark, since most are leaf-green in color and shaped like leaves. The scientific name of one of the more common genera, *Pterophylla*, is Greek for "leaf wing."

The two most common species in North America are rather similar in appearance. *Pterophylla camellifolia*, known variously as the true, rough-winged, or eastern katydid, is dark green with a brown triangle on its back where the relatively short rounded front wings overlap. The eastern katydid measures in at just under an inch and a half. The angular-winged katydid, *Microcentrum rhombifolium*, is slightly larger and can reach up to 2½ inches from head to wingtip. The wings are paler green and pointed instead of rounded.

Every fall, katydids lay flat oval eggs with a sword-shaped organ under loose bark or along the edges of twigs or leaves. The following spring the eggs hatch, and the immature katydids, or nymphs, feed on the leaves of many kinds of trees, including willow, poplar, elm, and cherry. They rarely occur in sufficient numbers to cause any economic damage. Come fall, the adult males, ready for mating, begin their nocturnal serenades to attract females.

In the process, they've attracted a lot of human attention as well. The call is generally rendered onomatopoetically as "katy-she-did, she did," and the identity of the Katy in question has given rise to innumerable stories. One account, written in verse by Mrs. A. C. Dufour in 1864, identifies Katy as one of two sisters vying for the heart of the same young gentleman—Oscar, as the story goes. Oscar made the singularly ill-advised decision to choose Blanche over Katy and mysteriously disappeared soon afterward. Hence the speculation repeated every night. It has been estimated that katydids can repeat their call as often as 60 times a minutes and up to 50 million times in a summer season. So if Katy didn't, after all that, she'll probably wish she had.

MARCH FLIES

Despite their name, March flies, or flies in the family Bibionidae, are actually more abundant in April or May than they are in March. Their name isn't really the only thing that's ill-fitting. For one thing, many March flies have tiny heads relative to the rest of their stocky little bodies, and in male March flies the eyes are so enlarged that virtually nothing else about the head can be readily distinguished. In contrast, female March flies have flattened heads with tiny eyes, which, according to F. R. Cole in *Flies of Western North America,* "give a grotesque reptilian appearance." Aside from their heads, March flies are more or less nondescript as flies go—small to moderate in size, very hairy, and generally dull-colored, although some species come equipped with a bright red or yellow thorax.

For the most part they're rather inoffensive, especially as flies go. The wormlike twelve-segmented larvae live mostly in and among plant roots, feeding on decaying organic matter, although they aren't beneath nibbling directly on plants on occasion; they've been accused of eating potatoes already damaged by wireworms, for example. Adults first

March fly: He fell in love with the wrong bug.

appear in early spring—hence "March" flies—at which time they don't win any awards for grace and beauty either. Cole writes further, "To the casual observer they seem to lack normal intelligence. On bright spring days they often appear in swarms, flying aimlessly or gathering on the blossoms of fruit trees, where they stumble about in a drunken sort of way."

For the most part, then, March flies are totally inoffensive. When they do make a nuisance of themselves, it really isn't by design. In parts of the southeastern U.S., around the Gulf of Mexico, for example, population densities in swamps are so high that they can literally stop traffic. So many March flies end up plastered on radiator fins that cars overheat and stall, and fly bodies litter windshields to the point that drivers can't see to drive. To add insult to injury, if the March flies aren't cleared off the car surface immediately, the finish can be permanently ruined. The worst offender in this regard is *Plecia nearctica*, whose traffic-stopping emergence is an annual event in the northern part of peninsular Florida. They're called lovebugs, since the vast majority that end up flattened on roads, radiators, and rights of way are in the process of mating—in flagrante delicto, as it were. With such distractions on their minds, they can almost be forgiven for failing to look both ways before crossing the street.

SNOWY TREE CRICKETS

Of all the right-wing propaganda on the air in summer, probably the easiest to listen to is produced by *Oecanthus fultoni* (*niveus*), the snowy tree cricket. Like most species in the family Gryllidae, snowy tree crickets chirp by rubbing their wings together. A thickened vein on the underside of one wing edged with ridges, the file, is drawn over a hardened scraper on the upper side of the other wing. In the case of the snowy tree cricket, the right wing is raised and drawn over the left. The right-wing messages being broadcast in this manner have nothing to do with a political viewpoint, however. The message is generally x-rated: snowy tree cricket males spend their evenings advertising their availability to prospective mates. If they're lucky, females, who have eardrums on their two elbow equivalents, orient to the sound for a rendezvous.

Each cricket species has a specific calling song with a distinctive chirp pattern and tone—the snowy tree cricket's is a bell-like sound. The basic pattern, though, is prone to local variation. The snowy tree cricket is distinctive in that, as the temperature increases, the number of chirps

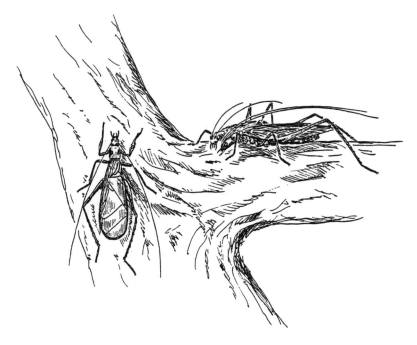

Snowy tree crickets: "Sure he's stridulating now, but will he respect you in the morning?"

increases; since the cricket happens to be calibrated to the Fahrenheit scale, you can estimate the temperature by counting chirps. The precise relationship was worked out by A. E. Dolbear in 1897, a physics professor at Tufts who published the now-classic paper entitled, "The Cricket as Thermometer." Professor Dolbear discovered that counting the number of chirps per minute, subtracting 40, dividing by 4, and adding 50 will give the approximate temperature in degrees Fahrenheit (for those who feel that getting up to check the thermometer doesn't present enough of a challenge). The formula has been immortalized as Dolbear's Law—although a subsequent academician, who recommended counting the number of chirps in 15 seconds and adding 37, failed to impress posterity enough to make his name in any enduring fashion.

Perverse entomologists have discovered that by heating up the male they can make him attract the wrong species of female. Females respond to faster chirps at higher temperatures. For example, the black-horned tree cricket female at 70° can pick the song of a black-horned tree cricket male out of a crowd when he's singing at 70,° but in the heat of the moment she'll also go for a male silvery tree cricket singing at 80° or a male four-spotted tree cricket at 89°.

Male snowy crickets don't stop at soft music when it comes to seduction. Once a female finds him, he lifts his wings and presents her with an aphrodisiac glandular secretion to put her in the mood. While she's busy eating, he makes his move. When it's all over and done, the male then breaks into a postcopulatory, or "staying together," song (called by the French the "triumphant song").

The rest of the snowy tree cricket's daily routine is far less conspicuous than its sex life. After mating, females lay their banana-shaped eggs in pinhole incisions in tree bark, sealing them in with a drop of excrement and an insulating secretion. The snowy tree cricket lays its eggs singly, the black-horned tree cricket in a long interlocking chain. The eggs hatch the following spring and the growing nymphs, whitish in color (whence cometh the "snowy"), eat anything they can get their mandibles into, including foliage, fungus, pollen, fruit, aphids, and other small insects. The adults, when not courting, are extremely reclusive and difficult to find. They reach about three-fourths of an inch in length and are greenish white in color; the males are approximately teardrop-shape in outline with large, flat, paddlelike wings.

Without their music-making wings, snowy tree crickets would probably be just another one of the undistinguished faceless insect hordes, but their bell-like song makes them unique. As Nathaniel Hawthorne wrote after hearing them go about their business, "If moonlight could be heard, it would sound like that."

YELLOWJACKETS

As autumn progresses and the temperatures fall, jackets begin to appear, and among the more conspicuous are yellowjackets. Yellowjackets and hornets are wasps in the family Vespidae. They're medium to large in size, sometimes surpassing an inch in length, and they're unmistakably marked either with black-and-white or black-and-yellow racing stripes. The color pattern is distinctive for a reason. Like black-and-yellow highway signs, the pattern serves as a memorable reminder for any potential enemy to "proceed with caution," since vespids are equipped with a very effective stinger at their posterior end. The front end of a yellowjacket from an insect point of view is no less imposing. While adult wasps use their long tongues for the pastoral pastime of sucking nectar and fruit juices, their strong mandibles (or jaws) are used to seize and masticate insect prey. Adult wasps don't eat the insects themselves but instead bring them back half-chewed to the nest and feed the pulp to developing wasp grubs there.

A vespid nest is about as recognizable as the wasps themselves. In the spring, fertilized female wasps (or queens) make a few hexagonal cells with a paperlike material produced by painstaking mastication of

Yellowjacket

wood chips, dried plant material, and the like. The cells are covered with a few thin sheets of the same paperlike material. The eggs hatch in about a week, and the queen feeds the developing grubs the insect equivalent of baby food until they pupate about two weeks later. The wasps emerging from these pupae differ from the queen in two ways: generally, they're smaller and they're sterile. These sterile "workers" spend the summer essentially as full-time domestic help for the queen, feeding the queen's new children and enlarging the paper nest until in some colonies it reaches a diameter of a foot or more.

This goes on amicably until fall, when the change in weather brings on a radical change in the colony. Fertile females are again produced and mate with the males or drones. At this point, the colony begins to fall apart. Old queens and drones die off and the workers scatter, abandoning the nest to fend for themselves. Things degenerate so much that the workers stop caring for the young altogether; they also have been known to start hauling the grubs out of the cells and eating them.

For insects, fall is not a particularly favorable season (flowers and fruits are in short supply), so every fall yellowjackets begin to hit garbage cans, picnic tables, dumpsters, and fast-food drive-in windows for scraps of stuff to sustain themselves. Their efforts are, for the most part, futile once winter hits and everybody dies off—everybody, that is, except the young queens, who have circumspectly crawled off to hide in crevices and cracks in and around trees in the woods until spring arrives, at which point they can start a new colony.

There's no real taxonomic difference between yellowjackets and hornets. Basically, the only major distinction is that hornets make their globeshaped paper nests aboveground, often suspended from trees or under eaves of houses; *Vespula maculata*, the baldfaced hornet, is an American representative of the aerial nest-makers. Yellowjackets, on the other hand, tend to nest in the ground in abandoned burrows, fallen treeholes, and similar cavities. Hornet nests become much more conspicuous in late fall mostly because the leaves begin to fall and leave the nests without effective camouflage. The camouflage doesn't really matter, since by the start of winter the nests are, for the most part, vacated. If you've ever been tempted to look inside a nest in winter, don't worry—nobody's home, except maybe for an overwintering queen or two. If someone tells you differently, you'll know it's a bald-faced lie.

Chapter 9

The Wet and Wild Bunch

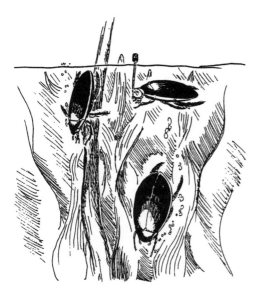

Backswimmers

Caddisflies

Diving beetles

Dobsonflies

Dragonflies and damselflies

Mayflies

Shore flies

Water striders

Whirligig beetles

BACKSWIMMERS

In economically depressed times, when it's hard to keep your head above water, take pity on *Notonecta undulata*, an insect that never seems to get its own head above water. *Notonecta undulata* is a backswimmer, a member of the family Notonectidae. Backswimmers spend most of their natural lives under the water, and, when they do come up for air, they do it tail end first. They can stay under water for long periods of time thanks to three longitudinal rows of hairs on their abdomen. These hairs form two compartments that trap and retain air for hours (although immature backswimmers who haven't mastered the technique have to come up for air every three to five minutes).

Backswimmers aren't the only aquatic true bugs (or hemipterans); they share slow-moving streams, ponds, lakes, and puddles with water boatmen, creeping water bugs, giant water bugs, waterscorpions, and water striders. They can easily be distinguished, however, from the madding crowds by their unique resting posture: they characteristically

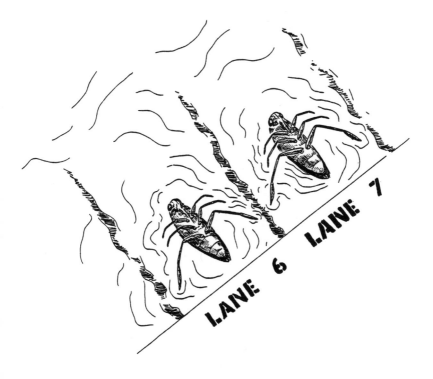

Backswimmers

hang at the water surface at a 45° angle, head down, tail up, and with their long fringed oarlike hind legs pointed downward toward their heads, poised (according to one author) like an oarsman waiting for the starting gun at the Yale-Harvard regatta. Backswimmers can also be recognized by their distinctive swimming style. As the name implies, backswimmers are masters of the backstroke. In fact, the wings on their back unite to form a ridge that acts like a keel when the bug flips over. What's more, backswimmers have reversed the usual protective color pattern; while most insects are dark on top and light underneath, backswimmers are white on top and black underneath.

Backswimmers remain active just about all year round and can often be spotted doing a leisurely backstroke in a hole in the ice covering a pond in midwinter. Mating takes place in early spring, and females lay elongate white eggs on vegetation or occasionally on the bodies of passing insects. The eggs hatch in about two weeks, and the nymphs go through five instars of grabbing and sucking their underwater neighbors. By July, they reach adulthood and leave the water, sometimes in swarms, to disperse, often ending up at lights at night.

Like most hemipterans, backswimmers have a beaklike mouth through which liquid food is sucked up. "Food" is anything that swims by, including tadpoles, small fish, water fleas, copepods, and other aquatic insects smaller than they are (which is smaller than, on average, a half-inch or so). Backswimmers generally store so much air, they have a natural tendency to float, which explains how they "hunt"— by clinging to vegetation, watching passing traffic, and letting go to drift upward when a tasty morsel appears. Once they secure their prey with their front and middle legs, backswimmers plunge their beaks into their hapless victims and pump in digestive enzymes that convert any standard-issue aquatic organism into soupy pulp. Backswimmers also are not averse to plunging their beaks into careless entomologists who handle them. While the bite isn't sufficient to convert the average entomologist into pulp, it is nevertheless painful and worthwhile going out of the way to avoid. After all, what could be more ignominious than a bit of back(swimmer)stabbing?

CADDISFLIES

For many people, October 31 is Halloween, a day for dressing up and hitting the streets in relentless pursuit of tricks or treats. But for caddisworms, October 31 means just another day of being all dressed up with no place to go. Caddisworms are caterpillar-like aquatic insects that construct an assortment of portable cases in which to house and protect themselves. The word "caddis" is derived from a Middle English term referring to bits and pieces of worsted yarn. Caddisworms are more likely to use stones, sand, pebbles, or twigs to construct their houses and use silk to hold it all together.

The cases are so distinctive that caddisworms can be recognized and identified by the case alone. For example, micro-caddisflies (in the family Hydroptilidae) make tiny cases like purses (only in their later instars) that average a quarter-inch in length. The Phryganeidae, or large caddisflies (the largest of the group), make cylindrical cases up to two inches long with sticks and leaves in a spiral arrangement. The Limnephilidae are called the log cabin caddisworms because of their habit of using sticks and leaves in crosswise construction to build a case, although some members of the family, marching to a different drummer, construct horn-shaped cases out of tiny bits of sand and shell painstakingly plastered together with strands of silk. And some psychomyiids make subterranean fallout shelters by lining burrows in the sand of stream bottoms with a cementlike substance. Then there are the philopotamids, which spin long fingerlike tubes out of silk. Most in keeping with the Halloween spirit, however, are the Helicopsychidae; these caddisworms arrange tiny sand grains into a helical case that's a dead ringer for a snail shell.

Caddisflies: "I love you, why won't you talk to me?"

Treats for growing caddisworms consist of algae for the most part, diatoms, and other aquatic vegetation. There are a few caddisworms, however, that are carnivorous. Some of these forms are free-living and, instead of making little houses out of silk, they spin nets which they anchor to vegetation and use to seine for passing insect and crustacean snacks.

Caddisworms grow slowly, often taking the better part of fall and winter to mature. When the casemaking caddisworms complete development, the case is closed with silk, and the caddisworms spin a cocoon and pupate inside. About a month later, the pupa (or rather the adult still inside its pupal case) chews itself out of the cocoon, swims up to the water's surface, and finds a rock, log, or comparable hard place on which to settle. Once settled, the adult emerges.

Adult caddisworms are called caddisflies, although properly speaking they aren't really flies at all. Caddisflies constitute their own order, the Trichoptera. The name, derived from Latin for "hairy wing," refers to their most distinguishing feature: the mothlike adults have two pairs of wings held rooflike over the body. Unlike moths, however, the wings are not covered with scales but are instead covered with hair. The *amount* of hair varies—some have a fine dense coat and other species are practically bald—but the hair is always there. The adults are far from flashy, tending toward shades of black, grey, or brown. By and large nocturnal, they often fly to lights at night. Left undisturbed, caddisflies mate and lay eggs in spring. Females can lay up to one thousand eggs, sometimes in gelatinous strings underwater or festooning overhanging rocks and bushes.

Since caddisfly adults and larvae make up the better part of the diet of many sport fish, including eastern brook trout, they are favorite models for fly-tying fishermen. Anglers tie dry flies to imitate the adults, and wet flies to imitate the larvae. For the fish who falls for one of these lures, it's evidently a "case" of mistaken identity.

DIVING BEETLES

Jacques Yves Cousteau usually gets credit for inventing the scuba (or self-contained underwater breathing apparatus) back in 1943. If the truth be told, though, credit probably belongs to the Dytiscidae, known worldwide as the diving beetles. Diving beetles are found in just about every conceivable kind of aquatic habitat and have developed some unique adaptations for an underwater lifestyle. Like most beetles, dytiscids have two hardened forewings called elytra. Unlike most beetles, however, the space under the elytra is specially designed for holding a bubble of air. When a diving beetle swimming underwater feels the need for a breath of fresh air, it swims up to the surface and projects the tip of its abdomen up above the surface film. Fresh air is then channeled directly into the underwing chamber and the beetle can dive back down again.

The diving beetle's self-contained underwater breathing apparatus is actually more sophisticated than Cousteau's scuba gear, because it can be partially recharged underwater. Oxygen, of course, is the element of importance for respiration, and it's present not only in the

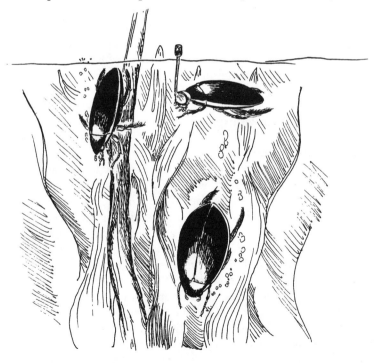

Diving beetles

diving beetle's air bubble, along with nitrogen and a few other gases, but also in dissolved form in the water all around. As the beetle uses up the oxygen in the bubble, dissolved oxygen from the surrounding water tends to move into the bubble to replace it. Eventually, however, the remaining gases in the bubble tend to diffuse out into the surrounding water and the bubble collapses, forcing the beetle back to the surface for more air. Since their life depends on getting back up to the surface every now and then for a recharge, diving beetles are very good swimmers. Like most aquatic beetles, their hind legs are flattened and equipped with long hairs for rowing; unlike most aquatic beetles, diving beetles stroke their hind legs together like oars, rather than alternately.

Larval diving beetles lack wings and therefore necessarily lack an underwing air chamber. Different species have different approaches to breathing underwater. While some must break the surface film periodically with the tip of the abdomen to gulp air, which is then stored in long internal trunk lines, others tap the air bubbles that form underwater around photosynthesizing plants, and still others make exclusive use of dissolved oxygen in the water by means of long feathery lateral gills on the abdomen. At the other end, larvae are equipped with long sickle-shaped hollow mandibles which they plunge periodically into the bodies of other underwater life forms. Paralyzing venoms and digestive fluids are pumped into the victim, and, when the body contents are sufficiently liquified, they're sucked back up through the same multipurpose mouthparts. Adult diving beetles are also predaceous, but they have rather mundane chewing mouthparts and must content themselves with crudely tearing apart and masticating their prey, which on a good day can range from crustaceans and insects to fish, frogs, and tadpoles.

In North America, diving beetles are faced with the fact that lakes, ponds, and other bodies of water freeze over, making it extremely difficult to breathe by poking your abdomen up above the water. While some species use ice as an excuse to enter diapause, settling down into the mud and keeping respiration at a minimum until spring, more enterprising species remain active all winter by making use of the oxygen-containing air bubbles trapped under the ice.

With over four hundred species, the Dytiscidae is the largest family of aquatic beetles in North America and, not inappropriately, contains some of the largest species of aquatic beetles in North America. *Dytiscus verticalis*, for example, a black beetle with yellow margins, is about 1¼ inches in length. If you think such a large juicy beetle would make a tasty morsel, think again; most dytiscids are well protected by chemical substances secreted by their pygidial gland in the tip of the abdomen.

These secretions contain such normal nasties as benzoic acid and benz-aldehyde but are relatively unique in the insect world in that they also contain steroids such as 11-deoxycorticosterone, testosterone, and es-tradiol. Some of these steroids are sex hormones in vertebrates. When a fish, amphibian, or mammal swallows a diving beetle, however, it responds not by getting amorous but by getting nauseated and vom-iting. Despite the sex hormones, then, dytiscids definitely don't make for a romantic meal.

DOBSONFLIES

Sportsmen have been known on occasion to raise hell—but fishermen's time would be more profitably spent raising *hell*grammites. A hellgrammite is the bizarre larval form of the equally bizarre dobsonfly, *Corydalus cornutus*. Dobsonflies, in the relatively obscure order Megaloptera, are among the most primitive insects with complete metamorphosis, in which larvae and adults look radically different and are separated by a pupal "resting" stage. Larvae and adults most assuredly look different in the dobsonflies. The six-legged greyish larvae, up to three inches long when full grown, are long and slender and equipped with large stout mandibles, a pair of abdominal claspers, and a set of tuftlike tracheal gills associated with each of the first seven or eight abdominal segments. These gills, along with a patch of water-repellent "hydrofuge" hairs on the ninth and tenth abdominal segments (to hold air underwater), allow the hellgrammite to pursue its aquatic lifestyle.

What hellgrammites actually spend most of their time pursuing is things to eat. They can and will eat anything they can sink their teeth into, including immature mayflies, caddisflies, black flies, midges, and even an occasional fellow dobsonfly. They feed mostly at night and

Dobsonfly (in larval stage, as hellgrammite)

can swim both forward and backward in snakelike fashion, although they seem to prefer the more insectan practice of crawling around in mud on stream bottoms. The larval stage is thought to take as long as three years. When it is completed, the hellgrammites haul themselves out of the water onto the stream bank, where they burrow into the ground or into logs to form a chamber in which to pupate. Pupation may take place as much as thirty feet from the water. Usually pupation is a quiet restful time for insects—not so for some dobsonflies, the pupae of which possess legs and mandibles so functional that they can be used to dispatch any other dobsonfly pupa intent on sharing a pupation chamber.

Things get stranger when, in about two weeks, the adult emerges from the pupal chamber. The adult dobsonfly looks like everybody's worst insect nightmare. First of all, they're enormous—up to *five inches* from the tip of the mandibles to the tip of the abdomen. Second of all, in the male about one-third of those five inches consists of sickle-shaped, deadly-looking mandibles. Appearances, however, can be deceiving, and the general consensus is that, since adult dobsonflies don't appear to feed, the mandibles in the male aren't used for biting anything. Their function is thought to involve courtship, possibly to be used for tenderly grasping the rather hefty female across the midsection (hence their large size). After mating, females lay literally thousands of eggs on plants, rocks, the undersides of bridges, or any other objects that dangle over water. In about four to five days, the eggs hatch and the larvae drop into the water to begin preying on the less fortunate.

Fishermen are quite familiar with dobsonflies and their kin, since they're unparalleled as live bait. Hellgrammites are taken by trout, bass, and most panfish. There's no such thing as a free lunch, of course, and using hellgrammites as bait is not without its disadvantages. Among other things, the stout strong mandibles can inflict a rather painful bite. So when you ask a fisherman, "How are they biting today?," make sure he knows you don't mean the bait.

DRAGONFLIES AND DAMSELFLIES

Probably more misinformation exists about dragonflies and damselflies than about any other group of insects. Just about everybody knows who they are—they're the long, thin, usually brightly colored insects that routinely buzz-bomb lakes, ponds, streams, stock ponds, and other bodies of water. Dragonflies are the larger, stouter ones with wings that broaden at the base and are held horizontally at rest. Damselflies (as the name suggests) are the delicate ones, usually metallic or shiny in color, with wings held vertically at rest. Both belong to the order Odonata.

While almost everybody can recognize them when they see them, not many people are exactly clear on what they do with their time—and here's where the misconceptions arise. Damselflies are sometimes called snake doctors, because they supposedly minister to sick and injured serpents, stitching up wounds and nursing them back to health. Some folks believe that snakes have to be completely pulverized in order to terminate them without interference from snake doctors. Dragonflies are also known as horse stingers or devil's darning needles, and they're supposed to sew up the eyes and ears of innocent children as they peacefully slumber. As far as modern entomology can ascertain, granting that there are limits to the science, there is no truth to these allegations.

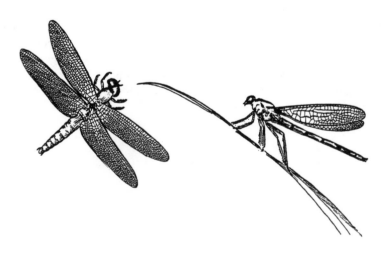

Dragonfly and damselfly: "That's a nice bit of stunt flying, but can you do it with *all* thirty thousand eyes closed?"

It's odd that such bizarre stories have arisen in connection with the odonates, since they're bizarre enough in their own right. Dragonflies and damselflies both begin life in the water as immatures called naiads. They live in muddy river bottoms on submerged plants or under debris and are enthusiastically carnivorous. Dragonfly naiads have elaborately modified mouthparts; the one that corresponds loosely to the lower lip is greatly elongated and jointed in several places. It folds under the head when not in use, but when a potential morsel swims by, the lip flips out, and two movable grappling hooks at the tip sink into the swimmer. They can even haul in small fish in this manner. In addition, both damselfly and dragonfly naiads breathe through gills. In damselflies, three leaflike projections on the tail serve as respiratory organs. In dragonflies, gills are located in the rectum, which doubles as a chamber for sucking in and forcing out water, creating forward motion by jet propulsion.

After several molts (up to fifteen for some species), taking anywhere from a few months to a few years, the naiads eventually climb up out of the water and shed their skin one final time and commence their adult lives. One common name for the adult dragonflies, mosquito-hawks, is certainly well earned—mosquitoes, gnats, and other flying insects are their favorite foods. The odonates in general are extremely effective predators. Their eyes, equipped with about thirty thousand individual facets, give them close to a 360°-field of vision, and their legs, equipped with interdigitating spines, dangle down in such a way as to form a sort of basket for scooping up insects on the wing—sort of the ultimate shopping cart.

Among the truly peculiar aspects of odonate life are mating arrangements. The male's genital opening is situated on the ninth abdominal segment, near the tip of the tail. The actual copulatory organs, however, are seven segments removed; they're on the abdomen two segments behind the last pair of legs. Before mating, the male has to deposit sperm from the tip of his tail into a storage sack behind his legs by bending his abdomen up and around. On the wing or on a plant, he then uses his tail to grab a female right behind her head. She then has to bend her body up under the male's and place her genital opening on the tip of her abdomen up against the storage sack right behind his legs. They can stay coupled in this position for several days, either to ensure successful fertilization or because they can't figure how to untangle themselves.

Things are even more bizarre among some damselflies. The male intromittent, or copulatory, organ has two spoon-shaped projections on its end, the function of which eluded entomologists for a long time. Evidently, the male uses these projections to scoop out any sperm left inside the female by an earlier consort before he deposits any of his own sperm, giving lie to the concept of "first come, first served" when it comes to fertilization of eggs.

If all of this talk about damsels and dragons sounds medieval, it's actually considerably more ancient than that. Fossil remains of dragonflies have been found dating back to the Permian period, 230 million years ago. Some of the prehistoric odonates had wing spans measuring up to 2½ feet. Imagine the nicknames people could have come up with for them.

MAYFLIES

According to entomologists, April showers bring not May *flowers* but rather *mayflies*. Mayflies are a familiar sight near any body of fresh water, including lakes, ponds, streams, and rivers. They're the fragile pale insects with triangular upright wings, long antennae, and a two- or three-inch-long tail projecting from the tip of their long delicate abdomen. Since they have four wings, not two, mayflies are not, properly speaking, flies—they're in their own order, the Ephemeroptera. Every spring (notably in May), mayflies make their presence known by forming tremendous swarms near the water, usually around dusk. These swarms, consisting almost exclusively of males, can number in the millions. They hang suspended in the air and move rhythmically up and down by the shore.

These swarms are actually mating aggregations. No long engagements here—females attracted to, or stumbling by, the swarm are almost immediately grabbed by a male for mating. Within the next forty-eight hours, females lay fifty to one thousand eggs on the surface of the water and die. The males die soon afterward, too. It's not for nothing that the name of the order is Ephemeroptera—the life of the adult mayfly is, if nothing else, ephemeral. They live long enough only to mate and lay eggs; the adults don't even take time to eat. Even if they had the time, they couldn't, seeing as they have no mouthparts and their guts are filled with nothing but air.

Mayflies are, on the whole, inoffensive animals. The immatures are called nymphs or naiads. Like the adults, they come equipped with a two- or three-part tail, but unlike the adults, they're anything but ephemeral. They spend up to a year or more living in the bottom of

Mayflies: "I don't know how to tell you this, but I'm just not ready for a long-term relationship."

lakes or streams, where they grub around for algae, plant material, and sometimes other insects. Some species molt up to twenty-seven times before maturing. When development is complete, the nymphs climb up a stem or up onto a rock. Once at the surface, the nymphs molt and turn into a subadult (or subimago), the only immature form in the entire Class Insecta equipped with wings. For several hours, up to a day or two, the subimagoes literally hang out to dry. Once their wings are dry, the subimagoes molt and finally turn into adults, for however briefly. Subimagoes can be distinguished from adults by their dull color and shorter tail, although one doesn't have a lot of time to try to figure out which is which.

Mayflies are important in freshwater ecosystems mostly because everything eats them—fish, birds, frogs, salamanders, dragonflies, and hordes of other organisms. For this reason, the mayflies are popular models for fishermen's flies; subimagoes are known as duns and adults as spinners. Mayflies are also important from the standpoint of ecosystem balance. They consume large amounts of decaying animal and plant material that would otherwise pollute the waters. They're very sensitive to changes in concentrations of dissolved oxygen in the water and so are used as indicator species; their disappearance from a lake or stream is an indication that the water quality is changing. However, despite their overall utility, their massive mating aggregations can be a nuisance. Since the males are attracted to lights, the swarms often drift toward towns. There, thousands of insects collect on windowsills, in doorways, and on streets, where they can cause accidents by creating slicks when tires run over them.

For the sake of thoroughness, it should be mentioned that in some parts of the country, mayflies appear earlier than in May. In these areas, they are sometimes referred to as shadflies, ostensibly since they provide sustenance for shad and other fish. In other parts of the country, they're variously known as lakeflies, dayflies, drakes, quills, sailors, and even cocktails. No doubt by the time you figure out what to call them, they'll be gone till next season.

SHORE FLIES

When *Ephydra bruesi* is up to its neck in hot water, it really couldn't be any happier. *Ephydra bruesi* and about twenty other species in the family Ephydridae can only be found in aquatic environments associated with the major geothermal regions of the world—in other words, in hot springs, geysers, and fumaroles, where water temperatures exceed 112°F on a regular basis. Although temperatures this high can hard-boil an egg, hot water doesn't much bother *E. bruesi*. The larvae—legless maggots equipped with few distinguishing features aside from terminal breathing tubes—spend their days grazing on the green and blue-green algae that constitute the sum total of plant life in hot water.

Generally, algae of hot springs form thick mats, either as mucilaginous sheets or silica-containing crusts. Mucilaginous algal mats are not necessarily an easy place to make a living. When a mat first forms, it's not suitable as food for ephydrids, because it's covered by hot water (the algae can withstand water temperatures 36° to 40° higher than

Shore flies: "Ah! . . . Hot tubs . . . I'm glad we moved to California."

197

the flies can tolerate). As the mat grows and thickens, the water is diverted, and the surface exposed to air cools to a more reasonable 110° or so. At that point, female flies move in to lay eggs.

However, after the eggs hatch, the larvae, as they feed, convert the algal mat into a thoroughly unappealing soupy mess. The weakened mat can no longer divert hot water flow, and eventually the mat collapses and hot water rushes in to parboil whatever larvae are left that haven't yet completed development. Adults, sometimes called shore flies, live by feeding on the dead bodies of insects which, lacking ephydrid fortitude, fall into hot springs and get cooked.

The family Ephydridae seems to thrive on adversity. A general trait of species in the subfamily Ephydrinae is a tolerance for water with high salt or mineral content. A number of these species, called brine flies, live in the effluent of salt mines or in the briny waters of salt lakes. Others live in the egg masses of spiders and frogs or in the dead bodies or excrement of animals, and still others mine the leaves of aquatic plants. One of these species, *Hydrellia griseola*, is actually a pest of rice crops. These and related underwater plant feeders obtain oxygen by piercing the air-filled stems of aquatic plants with a pair of spines at the tip of the abdomen.

Very little if anything about the appearance of the adults gives any indication of their unusual life-styles. Adults are small, fragile insects, usually less than one-tenth of an inch in length and generally undistinguished in color, although some species do have a metallic glint to them. The success of the group in unusual environments may be largely attributable to the youngsters. First of all, the retractible breathing tube allows them to live in material that's low in oxygen (such as anything decaying), since the air tube can extend for a considerable distance. Many of the larvae also have abdominal prolegs (or false legs), equipped with clawlike spines, and scalelike spines along their backs. All of these structures may prevent the larvae from backsliding when trying to move through slimy substances or slippery substrates (like mucilaginous algal mats).

Probably the slipperiest substrate an ephydrid ever finds itself in is an oil or petroleum pool. Larvae of *Helaeomyia petrolei* swim around in pure petroleum preying on the dead and dying insects trapped in the surface layer of the pools. It's not a very sporting proposition, preying on insects that are immobilized and can neither struggle nor fly away, but nobody ever denied that there's always been a lot about the oil business that is crude. . . .

WATER STRIDERS

While most of us go out of our way to avoid tension, there are insects that thrive on tension, and in fact couldn't live without it. Water striders, or species in the family Gerridae, for example, spend the better part of their lives skating along the surface tension of lakes, ponds, creeks, streams, and other bodies of water. The gerrids have a few tricks to help keep themselves afloat. First, there's a layer of non-wettable hairs on their tarsi (or feet) and on the lower half of their hind legs; since these are the only body parts in contact with the water, the surface film stays intact, because the water molecules stay more attracted to each other than to the water-repellent legs of the gerrid. Second, the claws that are usually found on the ends of insect legs in general are set far up on the legs of water striders, to minimize the risk of puncturing the surface film.

Water striders move across the water surface essentially by rowing, with their outstretched, greatly elongated middle and hind legs serving as oars. The much shorter front legs are held together under the head for support, although they are from time to time lifted out of the water to snag other aquatic insects as they come to the surface for a breath of fresh air, or to catch terrestrial nonwaterproof species that fall or land and get caught in the surface film. Like all of their relatives in the order Hemiptera, water striders use piercing mouthparts to puncture their dinner and suck it in.

Water strider: "If I thought about how I do this for very long, I know I'd fall in!"

Because of their unique ability to walk on water, water striders are known as, among other things, Jesus bugs or, less ecumenically, pond skaters. Unlike their human counterparts, however, they skate best when there's no ice on the pond. Consequently, in fall, some species of water striders fly to nearby wooded areas to spend the winter under rocks, logs, or litter. Following the spring thaw, they return to the water, where females lay their eggs, usually on objects that are floating, including bits of wood, feathers, seeds, shells, algae, and even an occasional insect, or at the water's edge. The newly hatched nymph heads right to the water surface and spends the next four to five weeks catching and eating whatever floats by, molting five times in the process.

Adults that emerge during the summer are often wingless; winged forms arise in the fall to head for the woods to hibernate. Even wingless gerrids can—and do—get around, however. Some species of *Halobates*, or ocean striders, live only in stretches of open sea and have been found over two hundred miles from land, despite the fact they never develop wings. Closer to home, the most commonly encountered species is *Gerris remigis*, a denizen of pools and streams throughout North America. About one-half to three-fourths of an inch in length, adults are most often wingless, black or brown on top, and silvery white underneath (due to the presence of a dense layer of waterproof hairs).

Water striders make use of the surface film not only as a means of support but also as a medium for communication. There are vibration sensors in the intersegmental membranes of the middle and hind legs which can detect approaching ripples. Sometimes the ripples indicate that a predator is approaching or that a potentially edible insect has landed nearby. Sometimes, however, water striders make waves of their own. A male interested in finding a female anchors onto a plant or a bit of floating debris and shakes his booty—that is, his rowing legs—at frequencies ranging from ten to thirty waves per second. A female that picks up the signal responds with low amplitude waves created by shaking her front legs up and down. The male answers in kind, and, once they rendezvous, the female indicates her interest by crossing one of her rowing legs over the male's leg or else by grabbing one of his middle or hind legs with her mouth. After mating is complete, the male backs off and generates more waves by shaking his rowing legs. Considering the nature of the messages being conveyed in the surface film, the film must by rights be rated "for adults only."

WHIRLIGIG BEETLES

If there are ever any Insect Olympic games, look for the gyrinid beetles to win all the gold in the aquatic events, since they can literally swim circles around anything in the water. Gyrinids are commonly known as whirligig beetles, and for once Latin and English names agree—if whirligig beetles have only one claim to fame, it's their habit of swimming in circles on the surfaces of ponds, lakes, and streams. Anatomically, they're well equipped for what they do. The shiny black or metallic green adults are oval and streamlined in shape. The middle and hind legs, bedecked with an outer fringe of hairs, are flattened into broad paddles for rowing, and the last exposed abdominal segment is long and laterally flexible to form a serviceable rudder. Around the outer edges of the body is a "plimsoll line," above which the body surface is water-repellent, so the beetle is essentially unsinkable.

As might be expected, the sensory equipment needed for controlling and coordinating this roundabout way of going is impressive in the extreme. For one thing, gyrinids effectively have four eyes. Each eye is divided into two parts on opposite sides of the head. The eyes on

Whirligig beetles: "This is great, just like bumper cars."

the top of their head project above the water to scan the skies for potential predators, while the lower two are submerged and search the waters for things to eat. Even their antennal structure is unique among beetles. The first two segments are shaped like a radar screen and the remaining six segments form a club. These antennae are held parallel to the water surface to detect the air currents above the waves in the water, which are created by the legs as they kick. These air currents provide directional information for the beetles to ensure that they travel only in the right circles.

Despite their name, whirligig beetles can get around in many ways. When aroused, they dive and swim with great facility. They are also eminently flightworthy and are known to fly two miles or more during the course of an evening. Adults also burrow in mud to overwinter. Early in the spring, they emerge from their burrows and congregate on water surfaces to find mates and get started with the business of reproducing. Females lay eggs in masses or in rows on underwater vegetation. The elongate larvae are equipped with pairs of feathery lateral gills on each abdominal segment, except for the last, which in addition to two pairs of gills has four hooklike structures. The larvae pass their time feeding on various and sundry aquatic life forms until they are ready to pupate, at which time they leave the water to construct a chamber out of whatever they can find along the water's edge. Adults emerge in mid or late summer, hang out a while scavenging insects off the water surface, and eventually overwinter underwater.

In the United States there are two common genera of gyrinids. Species in the genus *Gyrinus* are generally less than a quarter of an inch in length; species in the genus *Dineutus* are closer to a half an inch in length. The two genera can also be distinguished by smell. Adult gyrinids produce a milky secretion from glands in the thorax. These secretions ostensibly protect the beetles by repelling the advances of would-be predators. The odor produced by *Gyrinus* is repulsive, while the odor produced by *Dineutus* actually borders on pleasant. It's been likened to the aroma of ripe apples, and in fact the beetles are sometimes called "apple bugs." According to one source, in some parts of the country they're also known as "penny bugs," because if you put them under your pillow at night, there will be a "scent" there the next morning.

Chapter **10**

Pet Peeves

Black flies

Bot flies

Cattle grubs

Chicken lice

Ear mites

Fleas

No-See-Ums

Stable flies

Ticks

BLACK FLIES

If you have an interest in "current" events, then you ought to be familiar with simuliids, or black flies: black flies are in abundance wherever there's running water with a fast steady current. They're particularly common in springs in spring. In the water as larvae, or in the air as adults, black flies are hard to mistake for any other insect. The larvae or immature stages pass their time underwater attached to submerged vegetation, trailing roots, and smooth rock surfaces. They attach themselves by means of a circlet of hooks on their tail end. The circlet hooks into a silken pad produced by the salivary glands. The larvae have a similar circle of hooks on their single front leg, which projects out conspicuously underneath their heads. They can locomote in looplike fashion by alternately connecting and unconnecting the two circlets to the substrate. Their most conspicuous feature are the two cephalic fans, or mouth brushes, on either side of the head, which are used to filter flowing water for food particles. Growing black fly larvae aren't very particular, and, in addition to filtering and swallowing algae, protozoa, and other microscopic life forms, they occasionally intercept and ingest a younger, smaller, less fortunate fellow black fly.

Larvae undergo six to eight molts and pupate. As pupae, they spin a cocoon to undergo metamorphosis. After two or three days of pupation, the adult black fly emerges. Unlike the larvae, adult simuliids

Black flies

are completely terrestrial. The adults are often (but not invariably) black, stout, little flies scarcely more than a tenth of an inch in length. Their overall humpbacked profile with two stiff hornlike antennae conveys the overall image of a miniature buffalo; hence, yet another common name—buffalo gnats.

Males feed innocuously enough on nectar; for females, it's a different story. Although they are no less voracious than their offspring, they are a little more discriminating in what they eat. Unfortunately for humans who like to go camping, fishing, boating, or otherwise recreating in or near flowing streams, their meal of choice is the blood of warmblooded animals—almost any warmblooded animals. While no species feeds exclusively on humans, many like to vary their diet, and humans can be caught in the crossfire.

Unlike mosquitoes, which have neat syringelike bloodsucking mouthparts designed to pierce capillaries, black flies are "pool feeders"—that is, they lacerate the skin with their sawtoothed mandibles and then suck up blood from the oozing pool that results. The entire process is, as insect blood feeders go, rather slow and inefficient, and black flies can feed at a site for up to fifteen minutes or longer, taking up their own body weight or more in blood. Once at a puncture, they are determined feeders and can be dislodged only with difficulty. After they depart, residual salivary secretions in the wound can cause intense pain, itching, and swelling for some time afterward.

Most female black flies require a blood meal in order to develop and lay eggs, which they do with untempered enthusiasm on rocks, twigs, or vegetation at or just below the water's surface. Herein lies the problem. When adults emerge each spring, they do so in almost plague proportions, to the extent that there are records of domestic animals dying not from loss of blood but of suffocation, with nose, mouth, and breathing passages blocked with inadvertently inhaled black flies. Mercifully, black flies are extremely seasonal. Although they are exceedingly abundant each spring, by midsummer the overwhelming majority of adults are gone, and they cause few problems for the remainder of the summer—which is, of course, peak season for mosquitoes.

BOT FLIES

The life cycle of the bot fly is like something out of a horror movie, at least from a horse's point of view. Adult bots are innocuous at first consideration; they vaguely resemble bumble bees but have the courtesy to lack a stinger. They don't even have working mouthparts; not only do they not bite, they can't even eat. Basically, their sole purpose in life is to mate and lay eggs.

The three different species of bot flies that beleaguer horses have figured out three different ways to lay eggs. In the first species, the throat (or chin) bot flies, or *Gasterophilus nasalis*, females lay yellowish eggs on the hairs between the jawbones of the horse. The eggs hatch in five to six days and the tiny maggots crawl toward the chin and into the horse's mouth; once in the mouth they burrow down between molars, molt, and then are swallowed. Once in the horse's stomach they secure themselves to the stomach wall with grotesquely over-developed mouth hooks and spend the next nine months waiting for spring. That's when they let go, pass out with the manure, burrow in the ground to pupate, and emerge as adults. The females of the second species, *G. hemorrhoidalis* (nose, or red-tailed, bots), lay black eggs on the short hairs on the horse's muzzle. After two to four days the eggs hatch after coming in contact with moisture (as when the luckless horse licks his lips).

And last, but not least, is *G. intestinalis*, which has a different angle on parasitizing horses. The females lay their yellow eggs on hairs on the shoulder and forelegs. The eggs hatch when a horse has the misfortune of bringing its lips and warm breath in their proximity. The eggs hatch immediately in response to the temperature change, and the newly hatched larvae leap into the horse's mouth and begin to

Bot flies: "My goal in life is a Clydesdale."

burrow into its tongue. It takes about a month for the maggots to tunnel from the tip of the tongue to the back. Eventually, they emerge from the back of the tongue, are swallowed, and, like their cousins, sink their hooks into the stomach wall and wait for spring. Unlike the other two species, *G. intestinalis* attaches to the nonglandular part of the stomach. The other two species pass the winter in the glandular portion, on the other side of the tracks, as it were.

Granted, the bot fly's life-style is pretty revolting overall, but one must give bot flies their due. Theirs is not an easy life. For months they're stuck inside a horse stomach, which is not exactly Park Avenue. It's totally dark, and the maggots live surrounded by stomach juices more acidic than vinegar and at temperatures greater than 100°F. The free-living adults can neither eat nor drink and must find a horse on which to lay eggs before the first frost, since they're incapable of surviving cold temperatures. That's why, if you've got bot problems (or rather your horse does), fall is a good time to treat the horse. The adult flies are dead and can't reinfest horses that have been wormed.

There's a tremendous variety of insecticides available on the market, and the type of insecticide (as opposed to the brand) should be rotated at each worming to reduce the probability that resistance will develop. You should inspect the horse's legs for the yellow eggs; with practice they can be removed by combing or by scraping with sandpaper or killed by rubbing with kerosene or hot water. Be careful, though, not to parboil your horse in the process. Worm your horses in winter months to catch the maggots that are still residing in molars or tongues. By January, they should be anchored in the stomach and vulnerable to vermicides. Small wonder horses are such nervous animals—they have, not butterflies, but bot flies, in their stomachs.

CATTLE GRUBS

The expression "get off my back" may well have been invented by cattle acquainted with insects in the genus *Hypoderma*. *Hypoderma*, from the Greek for "under the skin," is known variously as the ox bot, ox warble, bomb fly, heel fly, or cattle grub, and it is one of the more grotesque insects in the neighborhood. The adult flies look innocent enough, though, as flies go. *H. bovis* is about three-fifths of an inch in length, *H. lineatum* scarcely one-half of an inch, and both species are covered with dense tufts of hair (*H. lineatum* with reddish orange hairs at the end of the abdomen, *H. bovis* with thick yellow hair on its thorax). Overall, both convey the impression of a natty bumble bee, rather than a common ordinary fly. But the trouble begins when females start to lay eggs in early spring.

Eggs are laid on the hairs of cattle around the heels or legs (hence the name heel fly). One to fifteen eggs are glued along a hair (and a female can lay up to eight hundred eggs). After about a week, the maggots hatch, crawl down to the skin, and bore into it either directly or through a hair follicle. The process takes over an hour and is extremely painful to the cow. Then, once under, the maggots take an extended tour of the body of the cow, crawling hither and yon, often stopping for a while near the esophagus. After about four months, they reach the spinal cord, burrow around for a short time, and finally cut through muscle to reach the back, where they molt.

Cattle grubs (heel flies)

After the molt, the tough spiny body of the maggot is sufficiently irritating to the body of the cow that a large swelling occurs around the maggot, filled with pus and fluid. This large swelling is called a warble. The maggot, unfazed by this reaction, merely cuts a small hole in the skin in order to breathe and happily feeds on the pus and fluids. After about a month, the larva molts again and gradually darkens from white to yellow to brown to black. When the process is complete, they exit through the breathing hole and crawl away to pupate. Left to their own devices, the grubs spend from one to three months inside the warble, passing the winter months with a steady supply of food and warmth. Altogether, the life cycle takes about a year.

Cattle grubs are bad news for everyone except another cattle grub. People raising cattle don't like them because they cost money. Milk production of infested cows can drop up to 25 percent; irritated beef cattle fail to put on weight; the exit holes of the grubs decrease the value of leather; and the infested area is unsaleable as meat, since the grubs convert firm red flesh to greenish yellow jelly. To add insult to injury the flies tend to prefer areas from which the best steaks are cut. Collectively, cattle grubs cause over $160 million in damage each year.

To say that *cattle* don't like cattle grubs is something of an understatement. When the flies are laying eggs in the spring, cattle go into a blind panic. As the flies try to reach the heels of a cow, it generally gallops around madly with its tail in the air, trying to reach shade or water, neither of which the fly will enter. Many cattle die by running into or over things in their desperate attempt to escape. The behavior can spread though an entire herd and cause stampeding. This mad panic is referred to as "gadding," and the term has entered our language with the expression "gadding about": "to be on the go with little purpose," according to *Webster's*. A "gadfly" is a person who irritates others. "Grubby" is another term that has come from the activity of cattle grubs; "grubby" hides aren't worth much money. Evidently, from a human perspective, the etymological contribution of the cattle grub far outweighs its entomological one.

CHICKEN LICE

The old line "there's nobody here but us chickens" is oftentimes *not* the case, an unfortunate situation from the chicken's point of view. Chickens, like most birds, are afflicted from time to time with insect parasites. Chief among these is the chicken body louse, *Menacanthus stramineus*. The chicken body louse, also known as the large poultry louse, lives in among the feathers and scales of chickens and turkeys and is found on guinea fowl, peacocks, pheasant, and quail who live with lousy chickens. Although it was once probably restricted to living on wild turkeys in North America, the chicken body louse is now worldwide in distribution, beleaguering chickens on virtually every continent. Chicken lice and their friends in the order Mallophaga chew

Chicken lice: "As a head louse in the chicken business, I've really had to claw my way to the top."

their food instead of suck it (unlike their relatives, the Anoplura, or sucking lice). They feed not on blood like the sucking lice, but on dry skin scales, feathers, scabs, and other delicacies. The chicken body louse *can* draw blood, though, by chewing through the skin and rupturing the pin feathers.

Even in winter all stages of the chicken louse can be found on infested chickens. The adults are approximately a tenth of an inch in length. They're totally wingless and are conspicuously flattened from back to front, the better to slip in and out among feathers. The three pairs of legs each terminate in two sharp claws, which rake the skin as the lice scurry all over the chicken and generally add to the discomfort of the infested fowl. Eggs are usually laid on the barbs of feather, especially near the vent or under the wings. The young nymphs, after hatching, look like their parents but are smaller and slightly yellow in color instead of brownish. After about ten to twelve days of eating skin scales and scabs, the lice are mature and ready to reproduce again. This phenomenal power of increase has unfortunate consequences for chickens. In one study, two people counted over thirty-five thousand lice on a single chicken and figured that they missed more than half that were present during the counting.

The chicken body louse isn't the only louse to louse up a chicken's life: the chicken head louse can be found in among the feathers on the top of the head, nibbling at skin scales; the shaft louse lives on the feathers themselves; and the fluff louse can be found in the fluff around the vent. Besides these there's also the brown chicken louse, the large chicken louse, the wing louse, and a host of others—all told, over forty species. While lice rarely kill adult chickens outright, they are so irritating that chickens fret, lose sleep, stop eating, and generally lose their avian joie de vivre. Chicks infested by lousy mothers, however, can acquire so many lice that they suffer heavy mortality.

There are several approaches to the control of chicken lice. One time-honored (but time-consuming) technique is the pinch method—applying insecticidal dusts by hand at the rate of one pinch per tail, vent, breast, neck, head, each wing, and each thigh, and two pinches for the back. Chickens can also be dipped or dusted. A more modern approach is to apply the insecticide not to the chickens but to the roosts or coop; a variety of insecticides have proved suitable. Keeping louse populations in check, though, can be a difficult and time-consuming task. One shouldn't get discouraged, though; just because the lice are always getting "down in the mouth" doesn't mean you should.

EAR MITES

If you ever get the feeling that the commands you give your disobedient doggie go in one ear and out the other, you may want to consider that, without your help, *Otodectes cynotis* goes in one ear and stays there. *Otodectes cynotis* is otherwise known as the ear mite, specifically the ear mite of dogs, cats, ferrets, and foxes. *Otodectes cynotis* spends its entire life deep in the recesses of the ears of carnivores, where it feeds on dead tissue and ear debris. An infestation of ear mites, known as otoacariasis, is seldom if ever fatal for the host, but it can be exceedingly annoying and is associated with tenderness, greyish discharge in the ears, loss of appetite, and depression. Dogs and cats afflicted with this problem rub and scratch their ears and shake their heads frequently. In cats, rubbing and scratching can produce a hematoma (or blood-filled swelling) in the ear, which may require surgical correction. Heavy infestation may damage the inner ear, where the mammalian balance organs are housed; animals so affected may lose their sense of balance, turn in circles, and hold their heads to one side.

The life cycle of the ear mite is not radically different from the life cycle of its near relatives in the family Psoroptidae, which include such lovelies as mange mites of horses, goats, sheep, rabbits, and reindeer, and scab mites of sheep, cattle, and horses. There are five life stages—egg, larva, protonymph, deutonymph, and adult—and adult females fall into two classes, pubescent (or pre-mated) and ovigerous (or egg-laying). Even at full size, ear mites are barely visible to the naked eye. To the discerning naked eye, their most distinguishing feature is the

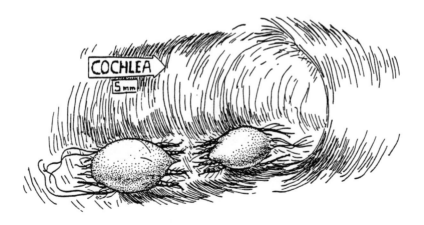

Ear mites

pair of long whiplike hairs (or setae) at the ends of the third and fourth pairs of legs. The ends of the first and second legs are equipped with strong claws, used for gripping and maintaining a leghold when an irritated host inserts a paw and begins digging and shaking.

One of the more unusual structures found on the mite is the pair of suckerlike organs on the tail end of the male. Males use these anal suckers for clasping onto pubescent females; during this period of attachment, the male drags the female around after him everywhere he goes. Apparently, nothing much goes on during this period. Mating takes place only after the female's final molt into the ovigerous stage. Since the mites are blind and not equipped to travel for long distances, attachment to the pubescent female appears to be the way a male guarantees he'll have a ready and willing mate some day. The attachment is so strong that it can continue even after a female molts or, worse yet, dies, so that the male travels everywhere with the dead body or empty shell of his erstwhile mate. It's the sort of relationship that can really be considered "kind of a drag."

FLEAS

The seemingly inevitable association between dogs and fleas led George Herbert, a poet and essayist of the seventeenth century, to remark pointedly that, "He who lies with dogs, riseth with fleas." It would probably have surprised Herbert no end to learn that most of the fleas on a given dog aren't *dog* fleas at all. That is to say, they're not necessarily *Ctenocephalides canis*, the dog flea—they could be *C. felis*, the so-called cat flea. Despite the name, the cat flea is also found on dogs, rats, squirrels, rabbits, chickens, and, most annoyingly, humans. Humans are afflicted with more than their share of fleas. Over twenty species are known to feed on humans, including *Pulex irritans*, the one privileged to earn the appellation, "human flea." Despite its name, *P. irritans* occurs also on cats, dogs, pigs, goats, rats, skunks, coyotes, ducks, and even spiny anteaters. In eastern North America, cat and dog fleas are the major offenders to human dignity.

Even though the menu may differ among fleas, the mode of eating is the same. All two thousand species of fleas feed as adults on blood. Fleas belong to the order Siphonaptera—from *aptera*, meaning "without

Fleas: "Did you hear what happened at the circus last night? The dog stole the show."

wings," and *siphon*, which describes the effective blood-letting devices with which they pierce skin and suck blood. Flea bites are distinctive in several ways. They generally are clustered in groups of two or three, they are often concentrated around the ankles and legs, and they invariably itch like you wouldn't believe.

The *aptera* in Siphonaptera sums up the situation fairly well—adult fleas lack wings and hence can't fly. They can, however, leap—a feat at which they truly excel. A flea can leap a distance 150 times its own body length, an accomplishment equivalent to a human doing a standing broad jump of a quarter-mile. In attaining a peak height of five or so inches in one two-thousandths of a second, fleas reach an acceleration of 140g, roughly equivalent to 20 times that required to put an Apollo moon rocket in orbit. Not all the force is muscular. Flea legs are equipped with resilin, an elastic protein that stores energy when it is compressed. When muscles relax to reduce the compression, resilin springs into action and puts the kick into the jump.

Fleas are protected from the extraordinary physical forces acting on their bodies during acceleration or landing by their tough hardened exoskeleton. This external armor also makes them virtually impervious to their major enemies—the prying, preening teeth, tongues, claws, and fingers of irritated hosts. Fleas are admirably suited for life aboard furry animals. They are flattened side to side so as to slip in among hairs without getting tangled, and they have backward-pointing comblike structures to anchor them in place if someone attempts to dislodge them before they're ready to debark.

Fleas begin life inside a tiny white egg which is dropped more or less randomly by the mother as she feeds. The eggs hatch in two to twelve days, depending on temperatures. The immature form of the flea is one of the better-kept secrets of the insect world. They're tiny, dirty white, wormlike creatures that live mostly in places cats and dogs sleep and where vacuum cleaners can't reach. There, they feast on shed skin, dust, and what's euphemistically called "flea dirt," the droppings of dried digested blood haphazardly deposited by their parents. Some fleas have remarkable methods for ensuring their offspring a nice nest in which to live. Female rabbit fleas, for example, only begin to develop eggs when the rabbit on which they are feeding gets pregnant. Flea and rabbit deliver at the same time and the baby fleas are comfortably ensconced in a nestful of baby bunny buddies.

Under the best of circumstances, larval development lasts nine to fifteen days. If meals are few and far between, fleas can take up to a year to develop. Mature larvae spin a rather sloppy, sticky cocoon to which dust and dirt adhere (to render them even more inconspicuous) and adults emerge anywhere from seven days to a year later. The pupae are extremely sensitive to vibration and are content to remain in their cocoons until passing footsteps herald the approach of an upcoming meal. This extraordinary patience is the reason that people moving into vacant or petless apartments might suffer a flea infestation soon after they sign the lease and move in. Adult fleas bide their time as well and can live three months or more without a meal.

Ridding one's house of fleas is a Sisyphean task unless one rids the house of the dog or cat. Since adult fleas spend 90 percent, and larval fleas 100 percent, of their time *off* the host, spraying or dipping your pet won't rid the house of fleas, although the dog or cat will certainly appreciate your efforts. Conscientious washing and vacuuming of floors and carpets (particularly where pets sleep) is effective to some extent. Several flea "bombs" or aerosal foggers now on the market contain methoprene, an insect growth hormone analogue that's hell on larval fleas. If the dog or cat regularly frequents the out-of-doors, treating the yard with insecticide is in order, too, since fleas lurk in vegetation waiting to pick up a meal.

Fleas are no fun. They spread such infirmities as dog heartworm, murine typhus, and, worst of all, bubonic plague. Cat and dog fleas are fairly poor agents of transmission for plague; the major vector is the Oriental rat flea, *Xenopsylla cheopsis*. If it's any consolation, fleas die of plague, too. Though bubonic plague is now to some extent amenable to antibiotic treatment, such was not always the case; in the "Black Death" of the fourteenth century, at least a quarter of the entire population of Europe was wiped out by plague. In the seventeenth-century plague, seventy thousand died in London alone. The poet George Herbert is known for yet another felicitously pithy phrase: "His bark is worse than his bite." While it might apply to dogs, it's certainly not the case for silent but deadly fleas.

NO-SEE-UMS

Whoever said that good things come in small packages obviously never encountered ceratopogonids. Ceratopogonids—variously known as biting midges, no-see-ums, punkies, gnats, or moose flies—do indeed come in small packages; the smallest ceratopogonid is only about a half a millimeter (about one-fifteenth of an inch) in length and can freely pass through screen doors. These very small insects, however, are capable of inflicting very nasty bites and causing pain out of all proportion to their size. The pain, however, as well as the swelling, blisters, and open sores that follow, is just the tip of the iceberg. Their less than desirable habit of feeding on the blood of horses, cattle, sheep, poultry, humans, lizards, and turtles and even the blood in the wing veins of dragonflies is outdone by their habit of transmitting a staggering array of pathogenic nematodes, viruses, and protozoa. Included in this rogue's gallery are bluetongue of sheep, horse sickness, onchocerciasis, and fistulous withers of horses, *Hepatocystis* of monkeys, *Parahaemopoteus*, *Leucocytozoon*, and *Akiba* of birds, and even a filarial parasite of frogs.

Until they sink their mandibles into your skin, ceratopogonids are easy to overlook. An adult resembles any of a number of types of little tiny flies, their most distinguishing feature perhaps being a Quasimodo-like hump on the thorax that projects over the head. Unlike mosquitoes, their wings are never scaled. Also, in contrast with mosquitoes, whose mouthparts work on the principle of a hypodermic syringe, ceratopogonid mandibles are scissorlike and cut, rather than painlessly pierce, the skin.

No-see-ums: "Look at that old guy! He must be nine hours old, if he's a minute!"

218

Like mosquitoes, however, only the female ceratopogonid sucks blood. She requires a blood meal to produce eggs. Fortunately, the amount of blood taken by even a very large ceratopogonid is small. *Culicoides variipennis*, for example, can only choke down about .6 mg at a time, only 20 percent of what a self-respecting mosquito would call a meal. Once she has mated and her eggs are fertilized, a female searches out a place to oviposit. Eggs and larvae are extremely prone to desiccation, so immature stages are basically aquatic or semiaquatic, inhabiting a variety of substrates including moist sandy ocean beaches, shores of streams, creeks or rivers, stock ponds, swamps, salt marshes, mud around water tanks, sewage effluent, holes in trees, and even inside pitcher plants. Many of these aquatic larvae—which are slender cylindrical wormlike creatures with a well-developed head and which may or may not have a single proleg (false leg) near the front end— are voracious predators in their own right, consuming aquatic insect eggs, nematodes, and other small life forms.

Culicoides brevitarsus, however, a pest of cattle, stays close to home and breeds in cattle dung, particularly in the wet spongy layers. The eggs of this species are equipped with a layer of hairs that physically trap air and permit respiration in the semiliquid manure. Development in the dung takes anywhere from three to four weeks, and the pupa prepares for adult emergence by anchoring itself into crevices in the dung pat by a pair of long spines at the tip of the abdomen.

An adult male's first order of business upon emergence is to find a mate, and to do so males of most species join a swarm. Depending on the species, a swarm may form either around a host or form independently of a host. Generally males face upwind and move up and down across any bit of distinctive ground cover that can serve as a marker. There's a peculiar sense of urgency associated with male ceratopogonids—a dispersed swarm, for example, re-forms after only ten seconds, and movement in and out of a swarm reminded one entomologist of "rush hour public transit." This may well be due to the fact that male fertility is a very fleeting thing; potency begins to decline at the ripe old age of 8 hours. Mating of a virile male only minutes old takes less than 8½ minutes but by the time he reaches 24 to 36 hours of age, a male can take over an hour to get the job done. Female ceratopogonids are hardly forgiving in matters of sexual performance. They are known to resist advances by older males by running away, tipping their abdomen, or even kicking violently. In some species, females enter a swarm to capture males and eat them and can occasionally be seen feeding on their own partner while in the very act—a consuming passion indeed.

STABLE FLIES

Despite what you might assume on hearing its name, the best place to find a stable fly is not inside a stable. Unlike its close relative, the house fly, the stable fly almost never lays its eggs on horse manure—or any other kind of manure for that matter. Rather, it lays its eggs on fermenting vegetation such as weeds, lawn cuttings, hay, or seaweeds. Granted, fermenting hay or straw is most likely to be found in or around stables, but stable flies are just as likely to be found breeding on the seaweed-strewn shores of lakes and oceans as they are at Belmont Park or Pimlico. To add to the confusion, the fly you're most likely to encounter in a *stable* is actually *Musca domestica*, which is, of course, the familiar house fly. Moreover, stable flies, or *Stomoxys calcitrans*, can occasionally be found in houses, where it's known as the "biting house fly."

Underlying this etymological confusion is entomological confusion as well. The stable fly is, to the uneducated eye, an exact double of the house fly. Both are greyish flies about one-fourth of a inch long with four dark longitudinal stripes on the thorax. A trained entomologist can recognize a stable fly by its slightly broader abdomen, its

Stable flies: "Here they come again. We're sitting ducks. I'd like to put the farmer in one of these things and see how he likes it."

habit of holding its wings widely spread at the tips instead of straight back, and its antennae, which are hairy only on the upper side. But even the rank amateur can detect the most important morphological difference between the stable fly and the house fly: while house flies have spongy soft mouthparts which they use to suck up liquified garbage of all descriptions, stable flies have bayonet-like mouthparts which they use to puncture skin and suck blood.

The stable fly isn't particular as to whom or what it punctures. *Stomoxys calcitrans* feeds readily on the blood of rats, rabbits, monkeys, cows, horses, goats, sheep, dogs, humans, and even birds and reptiles. Unlike most kinds of biting flies, in which only the females feed on blood, stable flies are affirmative action advocates—both male and female stable flies partake of blood readily.

Female stable flies lay their tiny whitish eggs, curved on one edge and straight on the other, in among spaces in loosely packed vegetation. One female can produce over six hundred in her lifetime. After two to five days, the eggs hatch into maggots, which basically resemble house fly maggots (distinguishing the two involves getting closer than most people care to do). Larval development takes about two to three weeks, after which time the larvae find a dry spot to molt into the brownish puparium. After one to two weeks, metamorphosis is complete, and an adult stable fly, anxious for its first blood meal, can emerge from its puparium in under an hour.

Although stable flies draw blood quickly, using their proboscis like an awl to puncture skin, and can feed to capacity in three to four minutes, they prefer to take a little at a time and change hosts frequently. Stable flies are therefore excellent vectors of disease and can carry the causative agents of trypanosomiasis on their mouthparts. They can cause weight loss and reduction in milk production in cattle not only by blood and tissue loss associated with blood feeding, but also by the constant aggravation and stress they create by persistent attack. So, in a pasture or paddock, *stable* flies mean *unstable* cattle or horses nearby.

TICKS

If you feel ticked off every summer, it's probably the fault of *Derma-centor variabilis*. *Dermacentor variabilis*, known as the American dog tick, is one of the commoner members of the bloodsucking family Ixodidae, otherwise known as the hard ticks. They're called hard ticks because the members of the family all share a shieldlike plate on their back called a scutum. In *Dermacentor variabilis* the scutum is white and, for a tick, fairly ornate. If for some reason you're ever interested in sexing American dog ticks, the male's scutum reaches almost down to its back end and puts a limit on the amount of expansion it can accomplish while feeding. In females the scutum reaches only a short distance behind the head (or capitulum), leaving them free to swell up to pea-size proportions while feeding.

Probably everyone is acquainted with the feeding habits of ticks—they suck the blood of just about any warm-blooded vertebrate they can sink their hypostome into. Adult *Dermacentor variabilis* are partial to dogs but won't pass up a tasty cow, horse, or human, given a chance. The immature stages, on the other hand, are fond of meadow mice and other smaller mammals.

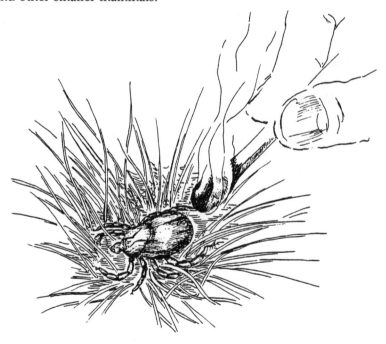

American dog tick

If you've ever wondered what happens after a tick feeds, if it's a female, she generally falls to the ground and in four to ten days lays from four to six thousand eggs. The eggs hatch after about a month, and the six-legged larva, or seed tick, looks for its first meal, often lying in wait on the tips of low-growing vegetation. After it finds a meal it molts into an eight-legged nymph, or yearling tick. The nymph repeats the whole process of finding a host and engorging. It then molts into an adult, finds a host, attaches for a week or two, and mates while feeding (think about that next time you check yourself for ticks). *Dermacentor variabilis* is a three-host tick, requiring a different individual host to complete each life stage. The related cattle tick, *Boophilus annulatus*, in the interests of efficiency is a one-host tick, taking all three meals and completing all three molts on a single host.

Since hosts are hard to come by, hard ticks are accustomed to hard times. Nymphs can live over a year without a meal, and adults can go two years between meals. Lest one feel sorry for a tick having to go hungry, one should keep in mind there's injury to the insult of tick bites. *Dermacentor* is an accomplished vector of Rocky Mountain spotted fever, Colorado tick fever, tularemia, bovine anaplasmosis, and possibly Lyme disease, and, even if it isn't spreading disease organisms, it can cause illnesses such as canine or tick paralysis by injecting salivary toxins.

Although a natural impulse when you find yourself parasitized is to grab the tick and yank it out, it's really not a great idea. Tick mouthparts are covered with sharp backward-pointing teeth that keep the whole assembly in place until the tick is ready to remove it. Dousing the tick with gasoline or alcohol or holding a smoldering match behind it will disturb it sufficiently to loosen its grasp, whereupon it can be removed with ease and dispatched. (Don't, though, douse the tick and ignite it). If you yank without preliminaries, the tick's head can remain in place while you dispose of the inoffensive body. In place, the head can cause infection or even paralysis. So a tick bite is no cause for panic—stay calm, don't lose your head, and, by all means, don't let the tick lose his.

Chapter 11

What's Eating You?

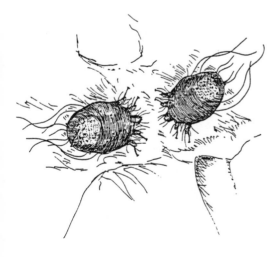

Bed bugs

Black widow spiders

Body lice

Brown recluse spiders

Chiggers

Crab lice

Human bot flies

Mosquitoes

Scabies

BED BUGS

There's something rather disconcerting about going to bed and waking with the realization that you're not alone. While on occasion this might be a pleasant surprise, such is definitely not the case when the unwanted guest is *Cimex lectularius*, the notorious bed bug. The bed bug lives on blood, and, while it doesn't *have* to be human blood (it's equally at home exsanguinating mice, chickens, rabbits, or even horses and cattle), the fact that it frequently *is* human blood is enough to have won the bed bug a rather unsavory reputation.

The bed bug is well suited to its profession; it feeds only at night, when its dinner's defenses are down. During the day, its flattened body shape allows it to conceal itself in cracks, crevices, and crannies in walls, ceilings, floors, or mattresses. The wings of a bed bug, reduced in order to facilitate slipping in and out of cracks, are nonfunctional as a result. Crawling, however, suffices for locating a meal. Only six blood meals are required to complete development—one before each

Bed bugs: "Traumatic insemination? That does it, I'm never getting married!"

227

molt and one before laying eggs—but each meal (particularly the later ones) can last up to ten to fifteen minutes. The bite itself isn't painful enough to rouse a person from deep slumber at the time, but, after the bed bug's long gone, the bite can cause intense pain and itching for up to a week. Bed bug bites are easily recognizable and distinguishable from bites of other nocturnal marauders by the fact that they are typically arranged in a systematic 1-2-3 lineup.

Females lay from one hundred to five hundred eggs, at the rate of about three or four per day, in furniture, behind baseboards, under wallpaper, and, in general, in any crevice large enough to accommodate a bed bug. The length of the developmental period varies with the availability of food and can range from a month or less to over a year. Actually, bed bugs are amazingly able to do without food altogether and are able to survive without eating for over a year, as one expects they might have to do in particularly seedy hotels. They can also make do with other warm-blooded creatures if humans are in short supply—house mice, for example.

The bed bug has been sharing sleeping quarters (although definitely not sleep) with humans quite literally since time immemorial. Historically, bed bugs are descended from forms that live on the blood of bats in caves. They are thought to have accidentally discovered humans back in the days when humans were still living in caves. The ancient Greeks were familiar (if not enamored) with the bed bug. Aristophanes, the Greek playwright, even mentioned them in two of his plays. They were widespread in Europe thoughout the Middle Ages, when the word "bug" was coined; originally it was intended for the bed bug exclusively. They arrived in North America, no doubt, with the early explorers and in fact have accompanied humans in their travels to the proverbial four corners of the earth. Bed bugs aren't quite the menace that they used to be, thanks in part to the extraordinary effectiveness of synthetic organic insecticides, but they still can hold their own in crowded prison camps, sleazy hotels, flophouses, brothels, and other dens of iniquity.

In view of the sort of living conditions it seems to prefer, it's not surprising that the sex life of the bed bug is somewhat less than refined. The female bed bug has no genital opening, so the male inseminates her simply by creating one himself. With his large sharp hooked intromittent organ, the male punches a hole in the body wall between the fifth and sixth abdominal segments of the female. There the female has an organ called the Organ of Berlese, which functions as a cushion to prevent the male's organ from doing permanent physical injury to her internal anatomy. The male bed bug then pumps in an enormous quantity of sperm, which swim into the bloodstream and eventually find the reproductive organs of the female. After the male withdraws his organ, the wound he created eventually closes and heals over, leaving a scar. The whole process is called, appropriately, traumatic insemination.

In fairness to the bed bug, it must be said that, of all the insects that feed on humans, the bed bug probably inflicts the least damage. The bite, while painful, is no more so than most, and, unlike fleas, mosquitoes, lice, or other blood feeders, the bed bug has never been implicated in the transmission of disease. About their most offensive trait is that they reek. The two pouchlike glands on their thorax produce an oily substance with an odor so obnoxious that the French word for bed bug, *punaise*, when used as an adjective, means "stinking." The same word can also be used as a noun to mean "thumbtack," for reasons not immediately apparent to this entomologist; entomology and etymology may on occasion be at odds.

BLACK WIDOW SPIDERS

Male chauvinism is probably nonexistent in *Latrodectus mactans*, known to most as the black widow spider. *Latrodectus mactans* is a member of the Theridiidae (or comb-footed spiders), and, like most spiders in the family, the adult females spin irregular free-form webs in which they hang belly-up and wait for prey. The black widow is easily but not always recognized by its glossy black or sepia body emblazoned on the underside with a crimson hourglass shape. The "hourglass" in some specimens can degenerate into a longitudinal stripe, a pair of blotches, or may even be absent altogether.

While most people have a vague notion as to what a female black widow looks like, only die-hard spider fans could give you a fair description of the males. Males are similar in size overall to females; at about 1⅛ of an inch in length (with legs extended), they're just slightly smaller than the females, at 1½ inches. They vary tremendously in color and range from near-black to mottled brown in color. About the only distinguishing features setting them apart from immature female black widows are a narrow abdomen and a pair of knobbed palps protruding from the head. The palps house the sexual apparatus, or intromittent organ, part of which bears an uncanny resemblance to a coiled watch spring.

There's a very good reason that male black widows don't even get billing in the common name of *Latrodectus mactans*. Female black wid-

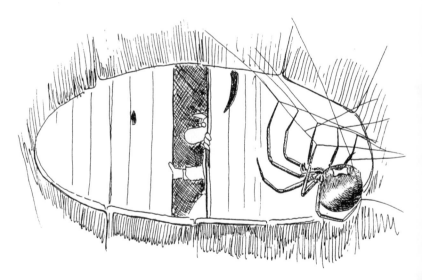

Black widow spider

ows possess a set of sacs which open into powerful fanglike chelicerae, or jaws. These sacs are filled with a potent neurotoxic venom, fifteen times more poisonous than the venom of the prairie rattlesnake (for those keeping score). While the venom is designed to immobilize and dissolve insects and other small organisms that blunder into their web, black widows are not averse to biting and injecting venom into larger blundering warm-blooded organisms. Since *Latrodectus mactans* is found in all forty-eight continental states in and around barns, stables, pump houses, woodpiles, sheds, garages, and houses, the blundering organism is often human.

The series of symptoms following the bite of a black widow is highly diagnostic. The bite itself, no more painful than a pinprick, is often not even felt. What follows, though, is intense pain at the site of the bite, agonizing pain in the groin, aching in the legs, and rigidity of large muscle groups, especially in the abdomen. Other symptoms include nausea, shock, headache, profuse sweating, and difficulty in breathing. Generally, the situation clears itself up in two to three excruciating days, although in at least 4 percent of cases death can result from respiratory paralysis. While the male possesses a set of venom glands, the mature male doesn't even use the venom to attack prey, and in fact the entire poison apparatus apparently becomes inactivated when the male matures. In a female, the glands measure about .4mm by 2.7mm; in males they're a paltry .16mm by .66mm, only one-sixteenth the size of the female's.

It's interesting to note that, in one ten-year study of arachnidism (or black widow bites), of thirty-seven people bitten, approximately 88 percent of them were males and approximately half of those were bitten on their private parts while using a privy. This may prove to be the ultimate form of female chauvinism.

BODY LICE

When it comes to lousing things up, nobody does it better than *Pediculus humanus*. *Pediculus humanus*, variously known as the head louse, body louse, grayback, cootie, seam squirrel, or motorized dandruff, is one of a select group of very few insects that make their living offending the health and dignity of human beings. Unlike its close relative, *Phthirus pubis*, the crab louse, which is more or less restricted to the coarse thick hairs of the private parts of humans, *Pediculus humanus* has diversified. For example, *P. humanus capitis*, the head louse, lives on the scalp in among the hairs of the head, whereas *P. humanus corporis*, the body louse, doesn't even live exclusively on the human body at all, residing instead in articles of clothing and only venturing forth from among the folds to hit the skin to feed—opting, apparently, for the "seamier" side of human existence.

The two subspecies, *P. humanus capitis* and *P. humanus corporis*, are virtually indistinguishable in appearance. Both are flattened, wingless, greyish, and tiny, although the body louse, at eight to sixteen one-hundredths of an inch in length, is slightly larger than the head louse,

Body lice: "I hate to be the one to break the news, guys, but that's not just hairspray."

a mere four to eight one-hundredths of an inch in length. Body lice tend to be slightly lighter in color, with more slender antennae and with less conspicuous abdominal segmentation—but for all practical purposes, the two subspecies are distinguished by their characteristic place of residence.

The life cycles of both the head and body louse are basically similar. Like all lice in the order Anoplura, they suck blood for sustenance, for which purposes they have three lancelike stylets for jabbing skin and letting blood; when not in use, these tuck away neatly into a little pouch. An immature molts three times before reaching adulthood, and each molt requires a blood meal. After about a month or so, the sexually mature lice mate and the females begin to lay eggs. Body lice drop eggs in among folds of clothing; head lice cement them to the base of the hair on top of the head. The eggs, or nits, are surprisingly large, about a millimeter long (almost one-quarter as long as the female louse).

After about a week to ten days, the eggs hatch. Since hair grows at a rate of about one-third of a millimeter per day, the eggs hatch about three millimeters away from the scalp. After the young nymph hatches from the egg, generally by swallowing air and bursting the egg open, the empty eggshell remains behind firmly glued to the still-growing hair. By the time the nits become conspicuous—as the hair gets longer— the lice are long gone, so nit-picking really *is* a useless enterprise.

Head lice are particularly partial to children, mostly because the only way head lice can get around is by direct contact between two heads, a frequent occurrence at schools and playgrounds. On girls nits are most frequent over or behind the ears, and on boys they're common near the top of the head. Parents, however, are by no means immune and may serve as an unsuspected source of an outbreak at a school.

Head lice really *can* louse things up. Among other things they can transmit such lovelies as epidemic typhus, relapsing fever, and trench fever. Physically removing them is exceedingly difficult. No force known to man can separate nit from hair near the scalp and crushing lice (an excellent way to infect oneself with any one of the aforementioned diseases, by the way) requires direct pressure of about five hundred thousand times the weight of the louse—if you can catch it in the first place. Although washing clothing in hot water effectively kills off body lice in clothing, plunging one's head in boiling water for ten minutes is not an effective alternative. About the only defense against head lice is regular inspection, brushing or combing (to dislodge stragglers), and, after lice are found, treatment with an insecticidal shampoo such as Proderm (Malathion lotion) or Kwell (with gamma-lindane). These kill both lice and nits and are, for the moment, the only effective means of sending head lice from "hair" to eternity.

BROWN RECLUSE SPIDERS

Bluegrass fans take note—there's one fiddle you'd be well advised *not* to pick up. The brown recluse, *Loxosceles reclusus*, is sometimes called the violin spider due to the presence of a dark violin-shaped mark on its otherwise undistinguished yellow to brown body. The female can be easily distinguished from the male by the fact that her fiddle has a much fancier peghead. The spider is most common in areas where bluegrass is popular—throughout the southern U.S., ranging from northern Georgia north to Indiana and southern Illinois, and south to Texas. The brown recluse is, in most respects, a very ordinary spider (for the purists, it has six eyes instead of the normal eight), and, like all spiders, it possesses chelicerae, hooklike digits, or "fangs," which contain the ducts of the poison glands.

Virtually all spiders have poison glands, which they use to paralyze or otherwise incapacitate insects and other small forms of life unfortunate enough to wander into a web or a spider's vicinity. But the brown recluse bears the dubious honor of being one of only two spiders in the U.S. whose venom is extremely poisonous to humans. The other,

Brown recluse spiders: "Hey, I didn't bite them! Must have been the malmignatte. . . ."

234

the black widow, *Latrodectus mactans*, already discussed, deservedly gets more press—of sixty-five spider-bite fatalities between 1950 and 1959, sixty-three were black widow bites—but a brown recluse bite is no fun either. However, *Loxosceles reclusus* is not terribly aggressive as spiders go, and in houses it tends to hide in corners, cracks, and clothing (one South American relative is called locally, *"arãna de detras de cuadros,"* or "spider of behind the pictures"). At only a half an inch or less in length, the spider is easily overlooked, particularly in dark corners.

Bites usually occur when a human inadvertently trespasses on a spider's turf. The bite itself is uneventful; except for a brief stinging sensation which may or may not occur, you might not even know you've been bitten. But in two to eight hours there's no mistake—then the pain begins in earnest as a thick weal or raised swollen area develops at the side of envenomation. As tissues die, a large necrotic lesion forms which, after a week or two, becomes dark and dry and separates from the rest of the skin to form an open ulcer that may take weeks to heal, depending on the size of the ulcer. In young children the bites are more serious, leading to death of red blood cells and subsequent hemolytic anemia and thrombocytopenia (a drastic reduction in the number of blood platelets). The medical terminology for brown recluse bites is, not inappropriately, necrotic arachnidism, or loxoscelism. While the condition is incapacitating, unsightly, and painful, it's rarely fatal, particularly in adults. It's some small consolation to know, too, that the brown recluse bites humans only in self-defense and not out of hunger or some sadistic sense of humor.

Loxosceles reclusus is not the only spider with musical affiliations. During the seventeenth century the town of Taranto, Italy, was seized with a peculiar misapprehension—that the bite of a spider called the *tarentula* (*not* tarantula) could be cured only by frenzied constant dancing to the point of collapse. In survivors, symptoms supposedly recurred on the anniversary of the bite. The whole episode was a product of mass hysteria instead of sound medical practice. First, *Lycosa tarentula*, a large conspicuous wolf spider, wasn't even involved in the fracas. The malmignatte, a black widow relative, is the only poisonous spider in the neighborhood. Secondly, dancing to the point of collapse has no known therapeutic value. Finally, in honor of the spider, the dance was called the tarentella, although it's unlikely that a spider, with eight legs, could ever manage anything more complicated than a two-step.

CHIGGERS

Chiggers, also called jiggers, redbugs, and a number of far more unflattering things, are actually arachnids rather than insects. They're mites in the order Acari. Their closest relatives, who are equally obnoxious, are the ticks. Mites, by and large, are anything but large—they range in size from minute to small, and chiggers are mid-sized as mites go.

Most people acquainted with chiggers have struck up the acquaintance by accident, walking through grass, weeds, or brambles in hot weather. In three to twenty-four hours, ankles, wrists, knees, waists, and any place where clothing contacts the body frequently begin to itch uncontrollably. The unrelieveable itching can persist for a week or longer and is accompanied in time by swelling, redness, pustules, and scabs. This summer delight is a chigger attack. If you aimed a microscope in among the pustules and scabs, you might be able to see a tiny bright red dot about 1/150 of an inch across. That's a chigger.

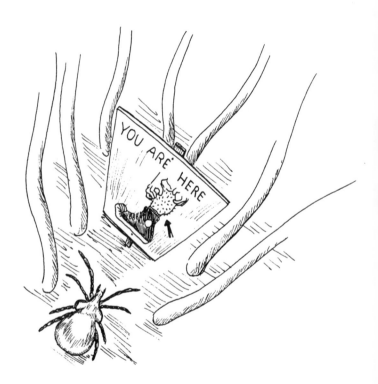

Chigger

Chiggers are the immature stages of several mite species in the family Trombiculidae. Every spring adult mites emerge from their overwintering sites underground and lay eggs. After about a week, the blind baby mite crawls out of the egg and climbs upward, preferably on tall vegetation. Unlike the adult mites, these first-stage larvae have only six (as opposed to eight) legs. They suffer from no great reduction in mobility as a consequence; they're extremely agile and can move very quickly. They're sensitive to carbon dioxide, because, since all animals breathe and in the process exhale it, carbon dioxide signals the approach of an impending meal. The larvae leap onto any passing vertebrate, and, lest humans feel too persecuted, it should be mentioned that they are equally attracted to chickens, ducks, frogs, toads, turtles, squirrels, cows, horses, rabbits, rodents, and a host of other animals. They're especially partial to snakes, on whom they settle down underneath the overlapping scales (and snakes, it should be said, have no appendages to scratch with either).

Once on their host, chiggers begin to feed. They don't, as many people believe, actually burrow down underneath the skin. Ewing in 1921 patiently (and courageously) observed twenty-six chiggers on his own skin and noticed that they attach to the skin surface or at the base of a hair. They don't actually suck blood, either. They simply inject an enzyme-filled fluid that disintegrates skin cells. This soupy mess of cytoplasm and cell bodies is sucked up by the mites. After they've eaten their fill, they drop to the ground and burrow into the soil. The swelling and itching are the result of an allergic response on the part of the host to the substances injected by the mites and can begin long after the chiggers have departed.

The immature mites remain in the soil and molt at least one more time before emerging as adults the next spring. Aside from the fact that they produce baby chiggers, the adults are completely harmless to other life forms. They mostly eat decaying wood and insect excrement.

Unfortunately, chiggers are not the only acarine occupants of human skin—the human body is a veritable zoo. Follicle mites, *Demodex folliculorum*, live in the hair follicles and sebaceous glands around the nose and eyelid. Follicle mites are worldwide in distribution and it's been estimated that three out of every four people harbor a population on their face. So to three-fourths of you readers out there—my consolations.

237

CRAB LICE

Although it's known in France, land of love and romance, as *papillon d'amour* or "butterfly of love," *Phthirus pubis* is hardly what Cupid had in mind for Valentine's Day. *Phthirus pubis*, better known to the American public as the pubic louse, or crab, is one of a very small number of very small insects parasitic on humans. A full-grown female crab louse measures about one-twelfth of an inch in length and one-eighteenth of an inch across; male crab lice are barely half the size of their mates. Like most parasitic insects, crab lice are characterized by a remarkable deficiency of seemingly necessary body parts. While they possess the full insect complement of six legs, they lack wings, and their antennae are greatly reduced. Most insects require antennae to locate food and depend on wings to get there; parasitic insects feed on hosts directly underfoot, eliminating the need for elaborate sensory systems and equipment for long-distance travel.

In the case of the crab, the need to travel is reduced still further by the accommodating nature of its human hosts. As its name implies, *Phthirus pubis* typically occupies the pubic region, where it firmly grasps

Crab lice: "You say you'll be working under cover tonight?"

pubic hair and inserts its barbed mouthparts into human flesh to suck blood. Crab lice disperse when human groins come in close proximity, although towels, toilet seats, and doorknobs are thought by some to mediate on occasion. With a nice warm place to live and three square meals a day close at hand (or tarsal claw), the crab louse's biggest problem is staying put. For this purpose, they are admirably well endowed. The width across the thorax of the crab measures about six one-hundredths of an inch, the approximate distance between hairs in the pubic region. In addition, the curvature and diameter of the tarsal claws on the second and third pairs of legs coincide nicely with the diameter of the relatively coarse hairs in the nether regions of human anatomy. To facilitate matters further, the claws are built on the principle of vise-grip pliers; a thumb-like hook projects outward and locks securely into a curved claw without muscular effort. It's such a tight grip that agitated hosts attempting to dislodge their uninvited lodgers often end up removing the louse and leaving the leg behind still gripping the hair. These formidable appendages are responsible for the common (if not popular) name of *Phthirus pubis*, the crab, or crab louse.

Among human parasites, crabs are more conservative than their relatives in the family Pediculidae. While the body louse, *Pediculus humanus*, can lay over two hundred eggs in a lifetime, the crab only lays a mere thirty or so. The eggs are cemented to a hair about eight one-hundredths of an inch from the skin. When the eggs hatch, the immature lice begin to feed immediately and continue to do so, stopping only to shed their skin three times on their way to adulthood. After their final molt, they have approximately one month in which to find a mate, lay eggs, and suck a little more blood before they expire.

As parasites go, crabs are despised way out of proportion to the damage they do. On a good day, only about three percent of the human population is infested with crabs. They have never been implicated in the spread of disease (unlike body lice and fleas), and, while their bites cause intense itching that is often socially embarrassing to relieve, they rarely move very far from their point of origin and can only switch hosts with human assistance. There is in fact probably no greater admirer of humankind. *Phthirus pubis* feeds on no other species and does not discriminate among the human races, feeding on all with great enthusiasm. While humans could no doubt willingly get along without the crab louse, without humans *Phthirus pubis* wouldn't have six legs to stand on.

HUMAN BOT FLIES

Dermatobia hominis really gets into people—to be specific, into their arms, heads, backs, abdomens, testicles, buttocks, thighs, armpits, and most other body parts. *Dermatobia hominis*, the human bot fly, is one of only a handful (or armpit- or thigh-full) of insects that actually develop inside people's bodies and feed on human flesh. Although they are not averse to attacking pigs, cows, cats, dogs, sheep, goats, horses, mules, monkeys, and even a few kinds of birds, *Dermatobia* makes a name for itself (at least the *hominis* part) by eating *Homo sapiens*, otherwise known as you and me.

Getting under a person's skin is no mean feat for *Dermatobia*. Anatomically, the insect isn't really equipped for piercing skin and injecting eggs. The adult fly, about one-half to three-fourths of an inch in length, resembles at first glance a run-of-the-mill bluebottle fly, to which it's fairly closely related. The adults are for the most part inoffensive, passing their time flitting about the forests of Mexico and Central and South America; some individuals may never lay compound eyes on a human being for their entire adult life.

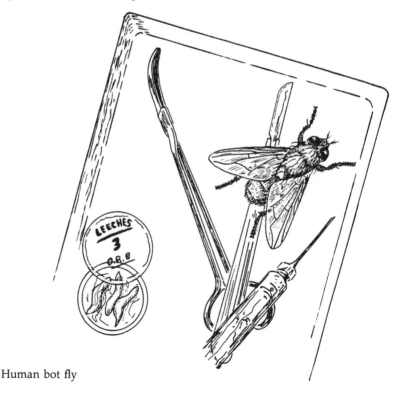

Human bot fly

However, the habits of *Dermatobia* eventually catch up to people. Female flies busy themselves with hunting female mosquitoes, black flies, deer flies, stable flies, and other insects that feed on human blood. When they catch one, they don't do the decent thing and kill it; rather, they snag it, often in mid-air, and lay eggs on its body, cementing them along the underside with a quick-drying adhesive substance. When the female fly or mosquito is released, it heads off to resume its normal business, to wit, searching for a likely source of a blood meal. Once the mosquito finds a human to her taste, she pierces the skin with her needlelike mouthparts and leisurely sucks her fill of blood. As she feeds, the warmth of the skin causes the bot eggs to hatch. The tiny maggots then penetrate the skin to snuggle into the subcutaneous tissue. A large, boil-like swelling (sometimes called a furuncle) forms around the flask-shaped larva as it completes development—a process that can take six weeks or longer.

Being a host or hostess to a human bot presents a number of operational problems. If, in the process of trying to extricate the larva, a person punctures or crushes it, maggot body parts or fluids can cause a life-threatening case of blood poisoning. The larvae are loathe to leave on their own volition, however. Some South American natives are said to lure the maggots out of their furuncles by placing a slab of bacon over the opening to the outside (the opening through which the maggot obtains oxygen). Left alone to mature, the maggot does little lasting injury and, at worst, causes itchiness and discomfort at night; one option, then (albeit an unpopular one), is to allow a maggot to complete its development and to use plural first-person pronouns until it crawls out on its own to drop to the ground and pupate.

Strange as it may seem, humans have on occasion actively recruited maggots to establish residence in their bodies. During World War I, an astute physician noticed that badly wounded soldiers left on the battlefield to die often recovered more quickly and completely than did soldiers who received medical treatment. Those with maggot-infested wounds fared particularly well. It turns out that larvae of various species of calliphorid blow flies selectively eat dead bacteria-laden tissue and thus cleanse and sterilize wounds. As a result, "surgical maggots," reared antiseptically in the laboratory, are still called into service on occasion to treat deep puncture wounds or bone infections. It has also been found that maggot infestation promotes faster healing as well due to the fact that the larvae excrete allantoin, a nitrogen-containing compound that kills bacteria and stimulates tissue growth. As far as surgical maggots are concerned, then, it's a case of "the ointment in the fly," instead of the other way around.

MOSQUITOES

Every spring the expression "What's eating you?" takes on an uncomfortable air of reality. In most parts of the U.S., spring is synonymous with mosquitoes.

Mosquitoes begin life in a surprisingly inoffensive manner. Immature mosquitoes, known as wrigglers, are completely aquatic. Equipped with chewing mouthparts, they spend the better part of their days feeding on bacteria, yeast, protozoans, and other bits of floating debris. After three larval molts, the immatures molt again and enter the pupal stage. Unlike most insect pupae, the comma-shaped mosquito pupae are very active and have earned the name "tumblers" for their manner of locomotion. After a few days (the number varies with the species) the pupa gulps air, splits down the middle, and produces an adult mosquito.

Male and female mosquitoes fly to flowers to obtain nectar—so far, no problem. And male and female mosquitoes court and mate in a number of picturesque ways. In some species, males (the ones with bushy antennae) form swarms into which a female must enter. In other species, males orient to certain sound frequencies, corresponding to those made by females. Oftentimes males accumulate in such numbers that they can clog machinery that happens to vibrate at the stimulatory frequency. This is a minor annoyance, granted. But after mating, the female mosquito requires a blood meal to produce eggs, and, again

Mosquitoes

depending on the species, as likely as not, the blood is supplied by a human being. Indeed, the mouthparts, or proboscis, of the female mosquito make up a surprisingly efficient insect imitation of a hypodermic syringe. There are six needlelike stylets; four actually cut the skin, and the remaining two are pressed together to form a groove through which blood of just about any description can be pumped.

If you really do want to know what's eating you, there are some easy ways of distinguishing the major types of mosquitoes, all of which require suppressing the urge to squash the insect in question to pulp on sight. *Anopheles* mosquitoes, members of which carry malaria throughout the world, generally lay their eggs in freshwater marshes, ponds, and the like. Each egg is equipped with an air sac (or float) to keep it near the surface. Adults generally have a covering of fine hairs on their bodies and have spotted wings. When they rest (taking a break from a heavy schedule of blood-letting), they hold their bodies pointed at an angle to the resting surface, butt end up in the air.

Aedes mosquitoes—these charmers can spread yellow fever and dengue, or breakbone, fever—tend to breed in rain pools, flood waters, and salt marshes, though some can breed in trees, pots, and the like. They often lay their eggs on dry soil which floods later; when the waters rise, the eggs hatch. *Aedes* adults rest with their body parallel to the resting surface, butt end pointing slightly downward. All are equipped with white scales on their bodies, and many have white bands encircling their legs and proboscis.

Finally, *Culex* mosquitoes, the ones that carry *Dirofilaria immitis* (dog heartworm), St. Louis encephalitis, both eastern and western encephalitis, and filariasis, breed in standing water in places like street gutters, polluted ponds, old tires, flower pots, and just about any kind of container that catches water. They've even been found breeding in baptismal fonts. Eggs are laid in "rafts" of several hundred eggs stacked in a single layer. The adults rest parallel to the surface, butt end parallel, and generally lack bands on both legs and proboscis. The most common *Culex*, *Culex pipiens*, the common house mosquito, actually doesn't even prefer human blood. Its tastes run more toward songbirds and poultry, but in a pinch it will feed on horses, dogs, cows, and people (not necessarily in that order). In northern climes, the males die at the onset of winter and the females hibernate in tree holes inside buildings, or in other protected places.

There are other more subtle ways of distinguishing among mosquitoes. The bites of *Aedes*, for example, tend to hurt more than the bites of *Culex* (hence the name of the common floodwater mosquito, *Aedes vexans*, which translates to mean, "disagreeable annoyance"). An alarming closing note—there are over 100 species of mosquitoes in

North America alone and over 2,400 worldwide. If you're thinking of traveling this summer to escape from mosquito torment, you'd better forget it—there are mosquitoes living from the steamy tropical rainforests to the polar ice caps, where they spend the winter frozen in ice. If you want to get away from them all, try heading for another solar system, just to be on the safe side.

SCABIES

The expression "I've got you under my skin" is usually a term of endearment. But when *Sarcoptes scabei* is under your skin, it's a different story altogether. *Sarcoptes scabei* is otherwise known as scabies, seven year itch, Norwegian itch, mange, and a number of shorter and less polite names. Scabies mites are tiny—at maturity females rarely top one-sixtieth of an inch in length, and males are only half that—but they wreak havoc out of all proportion to their size.

Scabies mites live just underneath the skin surface of humans and other animals. Actual infestation doesn't take very long at all. Adult mites, holding on with suckers on their front legs, prop up their hind end almost perpendicular to the skin surface, and, ripping and tearing with mouthparts and front legs, can completely disappear head first into the skin in under 2½ minutes. Once under the skin, female mites continue to burrow, leaving a sinuous, threadlike visible path up to an inch in length in their wake. Males tend to make much shorter burrows and in fact often just move into burrows abandoned by the more energetic females. Adults nibble on skin cells and interstitial fluids.

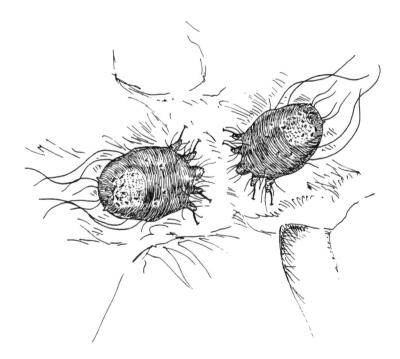

Scabies mite: "Give me some skin, man!"

Fertilized females begin to lay eggs only a few hours after mating and moving in, and they continue to lay two or three a day for up to two months. After three to four days, the eggs hatch and the hungry larvae wander over the skin surface looking for a nice place to settle in. They usually choose skin between the fingers, the bend of knee, elbow, or wrist, ankles and toes, underarms, breasts, and genitals. After two weeks of feeding on whatever they can find inside hair follicles, the sexually mature mites come out onto the skin surface to mate, and the whole process starts all over again.

The most annoying thing about scabies is that all this moving, shuffling, and mating is going on entirely unbeknownst to the human "mite motels." It takes up to six weeks for humans who have never had scabies to develop a sensitivity to the feces, the eggs, or the mites themselves. Then their troubles begin in earnest. A persistent itchy rash arises in affected areas. Scabies causes some of the most unbelievable itching known to humanity. Most of the complications of scabies arise from the patient's futile effort to relieve that itching. One diagnostic feature of a scabies infestation is that the unbearable itching gets even worse at night, when the mites are more active.

Fall and winter are prime seasons for scabies attacks. The mites are exceedingly transmissible and often reach epidemic proportions. The tendency of the mites to settle in pubic areas makes transmission by sexual contact possible, and scabies have acquired a somewhat sleazy reputation on that account. A respectable person can acquire an infestation, however, merely by shaking the hand of someone with scabies and it's even possible that contact with infested clothes or bedsheets can do the trick.

Despite its nickname, "seven year itch," people with scabies aren't doomed to almost a decade of scratching. Scabies, once diagnosed, is controlled relatively easily with insecticidal lotions containing lindane, available by prescription only. Once you get rid of scabies on your body, it's a good idea to launder clothing, bedding, or towels that may have come into contact with the mites. *Sarcoptes scabei* doesn't live very long off its hosts (two to three days tops) and sealing clothes into a plastic bag for a week or so will do the mites in nicely.

The four- to six-week delay in experiencing symptoms makes epidemiology a bit problematic for first-time sufferers. The place where you first begin to itch is almost certainly not the place where you contracted the infestation, so, if you vaguely remember friends complaining about a rash or itching about a month ago and—well, never mind, just scratch that last remark.

246

Select Bibliography

Many varied and sundry sources of information were used in assembling this book. Four main sourcebooks, consulted for every chapter, were constantly checked for physical descriptions, feeding habits, distribution records, and classification of the arthropods in question. These books were:

Borror, D. J., D. DeLong, and C. A. Triplehorn. 1976. *An Introduction to the Study of Insects.* 4th ed. New York: Holt, Rinehart and Winston.
James, M. T., and R. F. Harwood. 1969. *Herm's Medical Entomology.* New York: Macmillan.
Metcalf, C. L., W. P. Flint, and R. L. Metcalf. 1962. *Destructive and Useful Insects.* New York: McGraw-Hill.
Swann, L. A., and C. S. Papp. 1972. *The Common Insects of North America.* New York: Harper and Row.

Aside from these, I consulted various hard-core entomological books and journals for specifics of life cycles and for bizarre and unusual facts. The more important (and accessible) references are included in this bibliography. Other important sources of information are more difficult to document. I made heavy use of the University of Illinois Cooperative Extension Service entomology bulletins, which I receive weekly, and I thank the extension entomologists for unknowingly doing much of my legwork for me. Some information also came from news clippings and popular magazine articles. Yet another very valuable source of information has been the sum total of the Department of Entomology at the University of Illinois at Urbana-Champaign. My colleagues, expert in a wide variety of taxonomic groups, willingly shared literature references or first-hand knowledge whenever I called upon them. I would particularly like to thank Nathan Schiff and Steve Passoa in this regard.

So, with that said, what follows is a listing of the major references used in compiling each chapter.

Chapter 1

Benson, J. F. 1976. The biology of Lepidoptera infesting stored products, with special reference to population dynamics. *Biol. Rev.* 48:1–26.

Cornwell, P. B. 1976. *The Cockroach.* 2 vols. London: American Business Programmes.

Crowson, R. A. 1981. *The Biology of the Coleoptera.* New York: Academic Press.

Oldroyd, H. 1960. *Insects and Their World.* London: British Museum.

Oldroyd, H. 1964. *The Natural History of Flies.* London: Weidenfeld and Nicolson.

Sokoloff, A. 1972. *The Biology of Tribolium.* 2 vols. Oxford: Clarendon.

Sudd, J. H. 1982. Ants: foraging, nesting, brood behavior and polyethism. Chap. 2 in *Social Insects,* ed. H. R. Hermann, vol. 4, 107–55. New York Academic Press.

Taylor, R. 1975. *Butterflies in My Stomach or Insects in Human Nutrition.* Santa Barbara: Woodbridge Press.

USDA. 1979. *Stored-grain Insects.* Ag. Handbook 500. Washington, D.C.: USDA.

van Delden, W. 1982. The alcohol dehydrogenase polymorphism in *Drosophila melanogaster.* Selection at an enzyme locus. *Evol. Biol.* 11:187–222.

Wilson, E. O. 1974. *The Insect Societies.* Cambridge, Mass.: Harvard University Press.

Chapter 2

Davidson, R. H., and W. F. Lyon. 1979. *Insect Pests of Farm, Garden and Orchard.* New York: J. Wiley and Sons.

Hall, D. G. 1948. *The Blowflies of North America.* Baltimore: Thomas Say Foundation.

van der Lustgraaf, B. 1978. Ecological relationships between xerophilic fungi and house-dust mites (Acari: Pyroglyphidae). *Oecologia* 33:351–59.

Pfadt, R. E. 1985. *Fundamentals of Applied Entomology.* 4th ed. New York: Macmillan.

Untermeyer, L. 1959. *Golden Treasury of Poetry.* New York: Golden Press.

Vancassel, M., and M. Foraste. 1980. Le comportement parental des Dermaptères. *Reprod. Nutr. Develop.* 20:759–70.

Wilson, E. O. 1974. *The Insect Societies.* Cambridge, Mass.: Harvard University Press.

Chapter 3

Davidson, R. H., and W. F. Lyons. 1979. *Insect Pests of Farm, Garden and Orchard.* New York: J. Wiley and Sons.

Horsfield, D. 1978. Evidence for xylem feeding by *Philaenus spumarius* (L.)(Homoptera: Cercopidae). *Ent. Exp. Appl.* 24:95–99.

Jacques, H. E. 1951. *How to Know the Beetles.* Dubuque, Iowa: W. C. Brown Co.

Lederhouse, R. 1982. Territorial defense and lek behavior of the black swallowtail butterfly, *Papilio polyxenes. Behav. Ecol. Sociobiol.* 10:109–18.

Little, V. A. 1963. *General and Applied Entomology.* New York: Harper and Row.

Pfadt, R. 1985. *Fundamentals of Applied Entomology.* New York: Macmillan.

Self, L., F. Guthrie, and E. Hodgson. 1964. Adaptation of tobacco hornworms to the ingestion of nicotine. *J. Insect Physiol.* 12:224–30.

Tyler, H. 1975. *Swallowtails of North America.* Healdsburg, Cal.: Naturegraph.

Wiegert, R. G. 1964. Ingestion of xylem sap by meadow spittle bugs *Philaenus spumarius. Am. Midl. Nat.* 71:422–28.

Chapter 4

Beck, S. D., and J. Apple. 1961. Effects of temperature and photoperiod on voltinism of geographical populations of the European corn borer, *Pyrausta nubilalis. J. Econ. Ent.* 54:550–58.

Beck, S. D., and W. Hanes. 1960. Diapause in the European corn borer, *Pyrausta nubilalis* (Hueb.). *J. Insect Physiol.* 4:304–18.

Cowan, F. 1865. *Curious Facts in the History of Insects.* Philadelphia: J. B. Lippincott and Co.

DeLong, D. M. 1971. The bionomics of leafhoppers. *Ann. Rev. Ent.* 16:179-210.

Dixon, A. F. G. 1985. *Aphid Ecology.* Glasgow: Blackie.

Eisner, T., K. Hicks, M. Eisner, and D. Robson. 1978. "Wolf in sheep's clothing" strategy of a predaceous insect larva. *Science* 199:790–94.

Evans, H. E. 1978. *Life on a Little-Known Planet.* New York: E. P. Dutton.

Hunter, S. J. 1909. The greenbug and its enemies. *Bull. Univ. Kansas* 9 (#2).

Riley, C. V. 1877. *The Locust Plague in the United States.* Chicago: Rand McNally.

Chapter 5

Alcock, J. 1975. *Animal Behavior*. Sunderland: Sinauer Associates.

Brower, L. 1969. Ecological chemistry. *Scientific American* 220:22–29.

Brower, L, and J. V. Brower. 1964. Birds, butterflies and plant poisons: a study in ecological chemistry. *Zoologica* 49:137–59.

Clausen, L. 1954. *Insect Fact and Folklore*. New York: Macmillan.

Comstock, J. H., and A. B. Comstock. 1916. *How to Know the Butterflies*. New York: D. Appleton and Co.

Eisner, T., E. Tassell, and J. Carrel. 1967. Defensive use of a "fecal shield" by a beetle larva. *Science* 158:1471–73.

Foelix, R. R. 1982. *Biology of Spiders*. Cambridge, Mass.: Harvard University Press.

Gertsch, W. J. 1979. *American Spiders*. New York: Van Nostrand Reinhold Co.

Gorder, N. K. N., and J. Mertins. 1984. Life history of the parsnip webworm, *Depressaria pastinacella* (Lepidoptera: Oecophoridae) in central Iowa. *Ann. Ent. Soc. Amer.* 77:568–73.

Hodges, R. W. 1974. *Gelechioidea Oecophoridae. The Moths of North America North of Mexico*. Fascicle 6.2. London: E. W. Classey Ltd.

Holland, W. J. [1903] 1968. *The Moth Book*. New York: Doubleday, Page and Co., Dover Publications. (Contains quotations from Topsell, 1608.)

Opler, P., and G. Krizek. 1984. *Butterflies East of the Great Plains*. Baltimore: Johns Hopkins University Press.

Pyle, R. M. 1981. *The Audubon Society Field Guide to North American Butterflies*. New York: A. A. Knopf.

Roeder, K. D. 1935. An experimental analysis of the sexual behavior of the praying mantis (*Mantis religiosa* L.). *Biol. Bull.* 69:203–20.

Rutowski, R. L. 1978. The form and function of ascending flights in *Colias* butterflies. *Behav. Ecol. Sociobiol.* 3:163–72.

Rutowski, R. L. 1980. Male scent-producing structures in *Colias* butterflies. *J. Chem. Ecol.* 6:13–26.

Scudder, S. 1889. *The Butterflies of the Eastern United States and Canada*. 3 vols. Cambridge, Mass.: S. Scudder.

Silberglied, R., and O. Taylor. 1978. Ultraviolet reflection and its behavioral role in the courtship of the sulfur butterflies *Colias eurytheme* and *Colias philodice*. *Behav. Ecol. Sociobiol.* 3:203–43.

Weis, A., and W. G. Abrahamson. 1986. Evolution of host-plant manipulation by gall makers: ecological and genetic factors in the *Solidago-Eurosta* system. *Am. Nat.* 127:681–95.

Chapter 6

Bush, G. L. 1974. The mechanism of sympatric host race formation in the true fruit flies (Tephritidae). In *Genetic Mechanisms of Speciation in Insects*, ed. M. J. D. White, 3–23. Sydney: Australian and New Zealand Book Co.

Comstock, J., and A. Comstock. 1916. *How to Know the Butterflies.* New York: D. Appleton and Co.

Covell, C. V. 1984. *A Field Guide to the Moths of Eastern North America.* Boston: Houghton Mifflin Co.

Fisher, J. [1834] 1972. *Scripture Animals, a Natural History of the Living Creatures of the Bible.* Portland, Ore.: William Hyde; Princeton, N.J.: The Pyne Press.

Fitzgerald, T. 1976. Trail marking by larvae of the eastern tent caterpillar. *Science* 194:961–63.

Fitzgerald, T., and J. Edgerly. 1979. Specificity of trail markers of forest and eastern tent caterpillars. *J. Chem. Ecol.* 5:565–74.

Johnson, W. T. and H. Lyon. 1976. *Insects that Feed on Trees and Shrubs.* Ithaca, New York: Comstock Publishing Co.

Levine, E., and F. Hall. 1978. Pectinases and cellulases from plum curculio larvae: possible causes of apple and plum fruit abscission. *Ent. Exp. Appl.* 23:259–68.

McLain, D. K. 1979. Terrestrial trail-following by three species of predatory stinkbugs. *Fla. Ent.* 62:152–54.

McManus, M. L., and R. Zerillo. 1978. *The Gypsy Moth: an Illustrated Biography.* Washington, D.C.: USDA Home and Garden Bulletin 225.

Prokopy, R., and B. Roitberg. 1982. Foraging behavior of true fruit flies. *American Scientist* 72:41–49.

Scudder, S. 1889. *The Butterflies of Eastern United States and Canada.* Cambridge, Mass.: S. Scudder.

Waldbauer, G., and J. G. Sternburg. 1973. Polymorphic termination of diapause by cecropia: genetic and geographical aspects. *Bio. Bull.* 145:627–41.

Wallner, W. E., and L. Ellis. 1976. Olfactory detection of gypsy moth pheromone and egg masses by domestic canines. *Env. Ent.* 5:183–86.

Chapter 7

Alexander, R. D., and T. E. Moore. 1962. The evolutionary relationships of 17-year and 13-year cicadas and three new species. *Univ. Michigan Mus. Zool. Misc. Publ.* 121:1–58.

Buchler, E. R., T. Wright, and E. Brown. 1981. On the functions of stridulation by the passalid beetle *Odontotaenius disjunctus* (Coleoptera: Passalidae). *Anim. Behav.* 29:483–86.

Christenson, K. 1964. Bionomics of Collembola. *Ann. Rev. Ent.* 9:147–78.

Ford, N. 1926. On the behavior of Grylloblatta. *Can. Ent.* 58:66–70.

Henson, W. R. 1957. Temperature preference of *Grylloblatta campodeiformis*. *Nature* (1957): 637.

Reyes-Castillo, P., and M. Jarman. 1980. Some notes on larval stridulation in Neotropical Passalidae (Coleoptera: Lamellicornia). *Coleop. Bull.* 34:263–70.

Schuster, J. C. 1975. A comparative study of copulation in Passalidae (Coleoptera); new positions for beetles. *Coleop. Bull.* 29:75–81.

Schuster, J. C. 1983. Acoustical signals of passalid beetles: complex repertoires. *Fla. Ent.* 66:486–96.

Stannard, L. 1975. The distribution of periodical cicadas in Illinois. Illinois Natural History Survey Biological Notes, No. 91.

Sutton, S. 1980. *Woodlice.* New York: Pergamon Press.

Wheeler, W. M. 1935. *Demons of the Dust.* New York: Norton.

Chapter 8

Alcock, J. 1975. *Animal Behavior.* Sunderland: Sinauer Associates.

Clausen, L. 1954. *Insect Fact and Folklore.* New York: Macmillan.

Crowson, R. A. 1981. *The Biology of the Coleoptera.* New York: Academic Press.

Evans, H. E. 1978. *Life on a Little-Known Planet.* New York: Dutton.

Holt, V. M. [1885] 1978. *Why Not Eat Insects?* Faringdon: E. W. Classey Ltd. (reprint).

Kessel, E. L. 1955. The mating activity of balloon flies. *Syst. Zool.* 4:97–104.

Lloyd, J. E. 1965. Aggressive mimicry in *Photuris*: firefly femmes fatales. *Science* 149:633–34.

Lloyd, J. E. 1971. Bioluminescent communication in insects. *Ann. Rev. Ent.* 16:1–26.

Oldroyd, H. 1964. *The Natural History of Flies.* London: Weidenfeld and Nicolson.

Packard, A. S. 1887. *Half Hour Recreations with Insects.* Boston: Estes and Lauriat.

Waldbauer, G., J. Sternburg, and C. Maier. 1977. Phenological relationships of wasps, bumblebees, their mimics and insectivorous birds in an Illinois sand area. *Ecology* 58:583–91.

Wilson, E. O. 1974. *The Insect Societies.* Cambridge, Mass.: Harvard University Press.

Chapter 9

Brittain, J. E. 1982. Biology of mayflies. *Ann. Rev. Ent.* 27:119–47.

Cheng, L. 1973. The ocean strider *Halobates* (Heteroptera: Gerridae) in the Atlantic Ocean. *Oceanology* 13:564–70.

Cole, F. R. 1969. *Flies of Western North America*. Berkeley: University of California Press.

Corbet, P. 1963. *A Biology of Dragonflies*. Chicago: Quadrangle Books.

Corbet, P. 1980. Biology of Odonata. *Ann. Rev. Ent.* 25:189–217.

Clausen, L. 1954. *Insect Fact and Folklore*. New York: Macmillan.

Crowson, R. 1981. *The Biology of the Coleoptera*. New York: Academic Press.

Deonier, D., ed. 1979. *First Symposium of Systematics and Ecology of Ephydridae*. Erie, Pa.: North American Benthological Society.

Dettner, K. 1987. Chemosystematics and evolution of beetle chemical defenses. *Ann. Rev. Ent.* 32:17–48.

Evans, G. 1975. *The Life of Beetles*. New York: Hafner Press.

Mackay, R., and G. Wiggins. 1979. Ecological diversity in Trichoptera. *Ann. Rev. Ent.* 24:185–208.

McCafferty, W. 1981. *Aquatic Entomology*. Boston: Science Books Info.

Merrit, R. W., and K. Cummins, eds. 1978. *An Introduction to the Aquatic Insects of North America*. Dubuque, Iowa: Kendall/Hunt Publishing Co.

Milne, L., and M. Milne. 1978. Insects of the water surface. *Scientific American* 238:134–42.

Usinger, R. L., ed., 1956. *Aquatic Insects of California*. Berkeley: University of California Press.

Waage, J. 1979. Adaptive significance of postcopulatory guarding of mates and nonmates by male *Calopteryx maculata* (Odonata). *Behav. Ecol. Sociobiol.* 6:147–54.

Chapter 10

Baker, E. W., T. M. Evans, D. J. Gould, W. B. Hull, and H. L. Keegan. 1956. *A Manual of Parasitic Mites of Medical or Economic Importance*. New York: National Pest Control Association.

Greenberg, B. 1973. *Flies and Disease*. 2 vols. Princeton: Princeton University Press.

Horsfall, W. 1962. *Medical Entomology—Arthropods and Human Disease*. New York: Ronald Press.

Kettle, D. S. 1977. Biology and bionomics of bloodsucking ceratopogonids. *Ann. Rev. Ent.* 22:33–51.

Oldroyd, H. 1964. *The Natural History of Flies*. London: Weidenfeld and Nicolson.

Rothschild, M. 1975. Recent advances in our knowledge of the order Siphonaptera. *Ann. Rev. Ent.* 20:2451–59.

Smith, K. V. G., ed. 1973. *Insects and Other Arthropods of Medical Importance*. London: British Museum.

Soulsby, E. J. L. 1968. *Helminths, Arthropods and Protozoa of Domesticated Animals*. Baltimore: Williams and Williams.

Steelman, C. D. 1976. Parasites on domestic livestock production. *Ann. Rev. Ent.* 21:155–78.

Chapter 11

Andrews, M. 1984. *The Life that Lives on Man*. London: Arrow Books.

Berenbaum, M. 1980. Of life on man. *Cornell Engineer* 46:16–20.

Frazier, C., and F. Brown. 1980. *Insects and Allergy and What To Do About Them*. Norman: University of Oklahoma Press.

Harwood, R. F., and M. T. James. 1979. *Entomology in Human and Animal Health*. New York: Macmillan.

Horsfall, W. 1962. *Medical Entomology—Arthropods and Human Disease*. New York: Ronald Press.

Newson, H. D. 1977. Arthropod problems in recreation areas. *Ann. Rev. Ent.* 22:333–53.

Sasa, M. 1961. Biology of chiggers. *Ann. Rev. Ent.* 6:221–44.

Smith, K. G. V. 1973. *Insects and Other Arthropods of Medical Importance*. London: British Museum.

Zinsser, H. 1965. *Rats, Lice and History*. New York: Bantam Books.

Index by Scientific Name

Arthropods cited in the text are listed alphabetically along with their common name and taxonomic affiliations (family and order). For the most part, common names follow the publication, "Common Names of Insects and Related Organisms," assembled by the Committee on Common Names and published by the Entomological Society of America (College Park, Md.). This committee meets periodically and officially designates common names for use in the United States; although these names don't have the scientific status of Latin binomials, they do help to reduce confusion. Since the committee tends to focus on species of economic importance, some of the economically unimportant insects in this book don't have official common names. You can call them what you like in conversation until such time as the committee deems it appropriate to dub them.

As for pronunciation, while there may be rules, not many entomologists seem to honor them. "Common Insects of Kansas," a report of the State of Agriculture in Kansas published in 1943 by R. C. Smith, E. G. Kelly, G. A. Dean, H. R. Bryson, and R. I. Parker, has a pronunciation guide for family names for readers who enjoy puzzling out phonetic spelling.

Scientific Name	Common Name	Family	Order	Page
Acheta domesticus	House cricket	Gryllidae	Orthoptera	166
Acrosternum hilare	Green stink bug	Pentatomidae	Hemiptera	59
Aedes vexans	Vexans mosquito	Culicidae	Diptera	243
Agonopterix clemensella	No common name	Oecophoridae	Lepidoptera	105
Alaus oculatus	Eyed click beetle	Elateridae	Coleoptera	74
Anagasta kuhniella	Mediterranean flour moth	Pyralidae	Lepidoptera	14
Anasa tristis	Squash bug	Coreidae	Hemiptera	62, 63
Anopheles spp.	Malaria mosquito	Culicidae	Diptera	243
Anthrenus scrophulariae	Carpet beetle	Dermestidae	Coleoptera	28
Anthrenus verbasci	Varied carpet beetle	Dermestidae	Coleoptera	28
Aphidius testaceipes	No common name	Braconidae	Hymenoptera	85
Apis mellifera	Honey bee	Apidae	Hymenoptera	xiv, xv
Armadillidium vulgare	Pillbug	Armadillidiidae	Isopoda	155
Attagenus megatoma	Black carpet beetle	Dermestidae	Coleoptera	27, 28
Bathyplectes curculionis	No common name	Ichneumonidae	Hymenoptera	70
Battus philenor	Pipevine swallowtail	Papilionidae	Lepidoptera	136
Blatta orientalis	Oriental cockroach	Blattidae	Dictyoptera	4
Blattella germanica	German cockroach	Blattellidae	Dictyoptera	3
Bombyx mori	Silkworm	Bombycidae	Lepidoptera	82
Boophilus annulatus	Cattle tick	Ixodidae	Acari	223
Brachyrhinus ovatus	Strawberry root weevil	Curculionidae	Coleoptera	119
Calosoma calidum	Fiery hunter	Carabidae	Coleoptera	146
Calosoma scrutator	Fiery searcher	Carabidae	Coleoptera	146
Calosoma sycophanta	"Ground beetle"	Carabidae	Coleoptera	146
Camponotus pennsylvanicus	Black carpenter ant	Formicidae	Hymenoptera	25
Carpophilus dimidiatus	Corn sap beetle	Nitidulidae	Coleoptera	57

Scientific name	Common name	Family	Order	Page
Ceratina dupla	Little carpenter bee	Xylocopidae	Hymenoptera	164
Cerotoma trifurcata	Bean leaf beetle	Chrysomelidae	Coleoptera	47, 55
Chelymorpha cassidea	Argus tortoise beetle	Chrysomelidae	Coleoptera	98
Chrysopa oculata	Goldeneye lacewing	Chrysopidae	Neuroptera	86
Cimex lectularius	Bed bug	Cimicidae	Hemiptera	227
Colias eurytheme	Alfalfa caterpillar	Pieridae	Lepidoptera	109
Colias philodice	Clouded sulfur	Pieridae	Lepidoptera	109
Conotrachelus nenuphar	Plum curculio	Curculionidae	Coleoptera	132
Corydalus cornutus	Hellgrammite/dobsonfly	Corydalidae	Neuroptera	190
Ctenocephalides canis	Dog flea	Pulicidae	Siphonaptera	215
Ctenocephalides felis	Cat flea	Pulicidae	Siphonaptera	215
Culex pipiens	Northern house mosquito	Culicidae	Diptera	243
Culicoides brevitarsus	"Biting midge"	Ceratopogonidae	Diptera	219
Culicoides varipennis	"Biting midge"	Ceratopogonidae	Diptera	219
Danaus plexippus	Monarch butterfly	Danaidae	Lepidoptera	99
Demodex folliculorum	Follicle mite	Demodicidae	Acari	237
Depressaria pastinacella	Parsnip webworm	Oecophoridae	Lepidoptera	104
Dermacentor variabilis	American dog tick	Ixodidae	Acari	222, 223
Dermatobia hominis	Torsalo (human bot fly)	Cuterebridae	Diptera	240, 241
Dermatophagoides pteronyssinus	European house-dust mite	Epidermoptidae	Acari	39, 40
Diacrisia virginica	Yellow woollybear	Arctiidae	Lepidoptera	112
Dineutus spp.	Whirligig beetles	Gyrinidae	Coleoptera	202
Dissosteira carolina	Carolina grasshopper	Acrididae	Orthoptera	82
Drosophila melanogaster	Vinegar fly	Drosophilidae	Diptera	8
Dytiscus verticalis	"Diving beetle"	Dytiscidae	Coleoptera	188
Empoasca fabae	Potato leafhopper	Cicadellidae	Homoptera	88, 89
Ephydra bruesi	"Brine fly"	Ephydridae	Diptera	197

Scientific Name	Common Name	Family	Order	Page
Epicauta pennsylvanica	Black blister beetle	Meloidae	Coleoptera	72
Epicauta vittata	Striped blister beetle	Meloidae	Coleoptera	72
Epilachna borealis	Squash beetle	Coccinellidae	Coleoptera	54
Epilachna varivestis	Mexican bean beetle	Coccinellidae	Coleoptera	53, 54
Eristalis tenax	Drone fly	Syrphidae	Diptera	172
Estigmene acraea	Saltmarsh caterpillar	Arctiidae	Lepidoptera	112
Eurosta solidaginis	Goldenrod ball gall fly	Tephritidae	Diptera	95, 96
Forficula auricularia	European earwig	Forficulidae	Dermaptera	38
Gasterophilus haemorrhoidalis	Nose bot fly	Gasterophilidae	Diptera	207
Gasterophilus intestinalis	Horse bot fly	Gasterophilidae	Diptera	207, 208
Gasterophilus nasalis	Throat bot fly	Gasterophilidae	Diptera	207
Gerris remigis	Water strider	Gerridae	Hemiptera	200
Glischrochilus quadrisignatus	Sap beetle	Nitidulidae	Coleoptera	56
Glischrochilus sanguinolentus	Red sap beetle	Nitidulidae	Coleoptera	56
Grylloblatta	Rock crawler	Grylloblattidae	Grylloblattodea	152, 153
Gryllus spp.	Field crickets	Gryllidae	Orthoptera	166
Gyrinus spp.	Whirligig beetles	Gyrinidae	Coleoptera	202
Halobates spp.	Pelagic water striders	Gerridae	Hemiptera	200
Helaeomyia petrolei	No common name	Ephydridae	Diptera	198
Hilara sartor	Balloon fly	Empididae	Diptera	167
Hoplitis producta	Mason bee	Megachilidae	Hymenoptera	164
Hyalophora cecropia	Cecropia moth	Saturniidae	Lepidoptera	120, 121
Hydrellia griseola	Rice leaf miner	Ephydridae	Diptera	198
Hypera postica	Alfalfa weevil	Curculionidae	Coleoptera	69
Hypoderma bovis	Northern cattle grub	Oestridae	Diptera	209
Hypoderma lineatum	Common cattle grub	Oestridae	Diptera	209

Ips confusus (paraconfusus)	California five-spined ips	Coleoptera	Scolytidae	xiv
Isia isabella	Banded woollybear	Lepidoptera	Arctiidae	111, 112
Latrodectus mactans	Black widow spider	Araneae	Theridiidae	230, 231, 235
Lepisma saccharina	Silverfish	Thysanura	Lepismatidae	22
Limenitis archippus	Viceroy	Lepidoptera	Nymphalidae	100, 101
Loxosceles reclusus	Brown recluse spider	Araneae	Loxoscelidae	234, 235
Lycosa tarentula	Tarentula	Araneae	Lycosidae	235
Lymantria dispar	Gypsy moth	Lepidoptera	Lymantridae	127, 146
Lytta vesicatoria	Spanishfly	Coleoptera	Meloidae	72
Magicicada cassini	Periodical cicada	Homoptera	Cicadidae	150
Magicicada septendecim	Periodical cicada	Homoptera	Cicadidae	150
Magicicada septendecula	Periodical cicada	Homoptera	Cicadidae	150
Magicicada tredecassini	Periodical cicada	Homoptera	Cicadidae	150
Magicicada tredecim	Periodical cicada	Homoptera	Cicadidae	150
Magicicada tredecula	Periodical cicada	Homoptera	Cicadidae	150
Malacosoma americanum	Eastern test caterpillar	Lepidoptera	Lasiocampidae	123, 124
Malacosoma disstria	Forest tent caterpillar	Lepidoptera	Lasiocampidae	124
Manduca quinquemaculata	Tomato hornworm	Lepidoptera	Sphingidae	64, 65
Manduca sexta	Tobacco hornworm	Lepidoptera	Sphingidae	64, 65
Melanoplus bilituratus	Migratory grasshopper	Orthoptera	Acrididae	82
Melanoplus bivittatus	Two-striped grasshopper	Orthoptera	Acrididae	82
Melanoplus differentialis	Differential grasshopper	Orthoptera	Acrididae	82
Melanoplus femurrubrum	Red-legged grasshopper	Orthoptera	Acrididae	82
Melanoplus sanguinipes	Migratory grasshopper	Orthoptera	Acrididae	82
Melanoplus spretus	Rocky Mountain grasshopper	Orthoptera	Acrididae	83
Menacanthus stramineus	Chicken body louse	Mallophaga	Menoponidae	211
Metriona bicolor	Golden tortoise beetle	Coleoptera	Chrysomelidae	97
Microcentrum rhombifolium	Broad-winged katydid	Orthoptera	Tettigoniidae	174

Scientific Name	Common Name	Family	Order	Page
Microdon spp.	No common name	Syrphidae	Diptera	xvii
Misumena vatia	Crab spider	Thomisidae	Araneae	94
Monomorium pharaonis	Pharaoh ant	Formicidae	Hymenoptera	19
Murgantia histrionica	Harlequin bug	Pentatomidae	Hemiptera	58
Musca domestica	House fly	Muscidae	Diptera	xiv, 10, 220
Nezara viridula	Southern green stink bug	Pentatomidae	Hemiptera	59
Notonecta undulata	Backswimmer	Notonectidae	Hemiptera	183
Odontotaenius disjunctus	Horned passalus	Passalidae	Coleoptera	147
Oecanthus fultoni	Snowy tree cricket	Gryllidae	Orthoptera	177
Ostrinia nubilalis	European corn borer	Pyralidae	Lepidoptera	79
Otiorhynchus (Brachyrhinus) rugifrons	Eastern rough strawberry root weevil	Curculionidae	Coleoptera	119
Otiorhynchus rugosostriatus	Western rough strawberry root weevil	Curculionidae	Coleoptera	119
Otiorhynchus sulcatus	Black vine weevil	Curculionidae	Coleoptera	118, 119
Otodectes cynotis	Ear mite	Psoroptidae	Acari	213
Pachypsylla celtidisgemma	Hackberry bud gall maker	Psyllidae	Homoptera	131
Pachypsylla celtidismama	Hackberry nipple gall maker	Psyllidae	Homoptera	130
Pachypsylla celtidisvesicula	Hackberry blister gall maker	Psyllidae	Homoptera	130
Pachypsylla venusta	Hackberry twig gall maker	Psyllidae	Homoptera	131
Papilio glaucus	Tiger swallowtail	Papilionidae	Lepidoptera	135, 136
Papilio polyxenes	Black swallowtail	Papilionidae	Lepidoptera	136
Papilio troilus	Spicebush swallowtail	Papilionidae	Lepidoptera	136
Pediculus humanus capitis	Head louse	Pediculidae	Anoplura	232, 239, xiv
Pediculus humanus corporis	Body louse	Pediculidae	Anoplura	232, 239, xiv

Scientific name	Common name	Family	Order	Page
Periplaneta americana	American cockroach	Blattidae	Dictyoptera	4
Philaenus spumarius	Meadow spittlebug	Cercopidae	Homoptera	60
Photuris spp.	Firefly	Lampyridae	Coleoptera	170
Phthirus pubis	Crab louse	Pediculidae	Anoplura	232, 238, 239
Phyllophaga spp.	May (June) beetles	Scarabeidae	Coleoptera	51
Phylloxera vitifoliae	Grape phylloxera	Phylloxeridae	Homoptera	125
Phryxothrix spp.	"Firefly"	Lampyridae	Coleoptera	170
Plecia nearctica	"Lovebug"	Bibionidae	Diptera	176
Plodia interpunctella	Indian meal moth	Pyralidae	Lepidoptera	13
Pollenia rudis	Cluster fly	Calliphoridae	Diptera	33, 34
Porcellio laevis	Sowbug	Porcellionidae	Isopoda	155
Psychoda alternata	Trickling filter fly	Psychodidae	Diptera	18
Pterophylla camellifolia	Northern katydid	Tettigoniidae	Orthoptera	174
Pulex irritans	Human flea	Pulicidae	Siphonaptera	215
Reticulitermes flavipes	Eastern subterranean termite	Rhinotermitidae	Isoptera	42
Rhagoletis pomonella	Apple maggot	Tephritidae	Diptera	115
Rhopalosiphum maidis	Corn leaf aphid	Aphididae	Homoptera	77, 78
Rodolia cardinalis	Vedalia beetle	Coccinellidae	Coleoptera	53
Sarcoptes scabei	Itch mite	Sarcoptidae	Acari	245, 246
Schistocerca americana	American grasshopper	Acrididae	Orthoptera	82
Schizaphis graminum	Greenbug	Aphididae	Homoptera	84
Scolopendra spp.	Centipedes	Scolopendridae	Scolopendromorpha	29
Scutigera coleoptrata	House centipede	Scutigeridae	Scutigeromorpha	29
Sphecius speciosus	Cicada killer	Sphecidae	Hymenoptera	143, 144
Stagmomantis carolina	Carolina mantid	Mantidae	Dictyoptera	107
Stenopelmatus longispina	Jerusalem cricket	Stenopelmatidae	Orthoptera	173
Stomoxys calcitrans	Stable fly	Muscidae	Diptera	220, 221
Supella longipalpa	Brown-banded cockroach	Blattellidae	Dictyoptera	4

Scientific Name	Common Name	Family	Order	Page
Tenebrio molitor	Yellow mealworm	Tenebrionidae	Coleoptera	16
Tenebrio obscurus	Dark mealworm	Tenebrionidae	Coleoptera	16
Tenodera aridifolia sinensis	Chinese mantid	Mantidae	Dictyoptera	106
Theridion tarrenfossorum	Cobweb spider	Theridiidae	Araneae	36
Theridion tepidariorum	Cobweb spider	Theridiidae	Araneae	35
Thermobia domestica	Firebrat	Lepismatidae	Thysanura	22
Tinea pellionella	Casemaking clothes moth	Tineidae	Lepidoptera	31, 32
Tineola biselliella	Webbing clothes moth	Tineidae	Lepidoptera	31, 32
Torymus spp.	No common name	Torymidae	Hymenoptera	131
Tribolium castaneum	Red flour beetle	Tenebrionidae	Coleoptera	6
Tribolium confusum	Confused flour beetle	Tenebrionidae	Coleoptera	6
Tribolium giganteum	Giant flour beetle	Tenebrionidae	Coleoptera	6
Trichophaga tapetzella	Carpet moth	Tineidae	Lepidoptera	32
Trichopoda pennipes	Tachina fly	Tachinidae	Diptera	63
Vanessa cardui	Painted lady	Nymphalidae	Lepidoptera	102
Vermileo spp.	Wormlions	Rhagionidae	Diptera	142
Vespula maculata	Baldfaced hornet	Vespidae	Hymenoptera	180
Xenopsylla cheopsis	Oriental rat flea	Pulicidae	Siphonaptera	217
Xylocopa virginica	Carpenter bee	Anthophoridae	Hymenoptera	163, 164

A Note on the Author

MAY R. BERENBAUM is associate professor of entomology at the University of Illinois at Urbana-Champaign. She first began to appreciate arthropods as an undergraduate biology major at Yale University where she learned the important distinction between insects that bite people and insects that don't. After graduating from Yale *summa cum laude* with honors in biology in 1975, Berenbaum entered graduate school at Cornell University in ecology and evolutionary biology and received her Ph.D. in 1980. Her thesis research focused on the biology and chemistry of plants in the Umbelliferae (carrot family) and the insects associated with them. Her research has been recognized by the National Science Foundation, which in 1984 made her one of two hundred Presidential Young Investigators nationwide. She also received a Guggenheim Foundation fellowship for her studies of the chemical mediation of plant-insect interactions. Some of the insects she studies (notably the black swallowtail, the tiger swallowtail, and the parsnip webworm) are featured in this book.

In addition to research, Berenbaum is involved in teaching entomology and other biological sciences. Her interest in acquainting the general public with insects has resulted in a number of popular science publications and frequent radio and television appearances; in addition she organizes the annual Insect Fear Film Festival on the University of Illinois campus. She is currently working on a sequel to *Ninety-nine Gnats, Nits, and Nibblers*, featuring ninety-nine more amazing arthropods.

A Note on the Illustrator

JOHN PARKER SHERROD is a technical assistant at the Sections of Economic Entomology and Faunistics of the Natural History Survey. John received a B.S. in Medical Art and Biocommunications at the University of Illinois at Urbana-Champaign. Among his many artistic accomplishments, he is illustrating the Peterson's Field Guide to Freshwater Fishes of North America. He is a member of the Guild of Natural Science Illustrators and is working on several children's books.